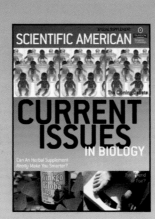

On the cover:
top photo: © Victor de Schwanberg/Photo Researchers, Inc.; bottom left: © Cordelia Molloy/Photo Researchers, Inc. ; bottom right: © Jpl / Anne/Photo Researchers, Inc.

Current Issues in Biology
is published by Scientific American, Inc. with project management by:

Jeremy Abbate, PROJECT DIRECTOR, MANAGER OF CUSTOM PUBLISHING

Bryan Christie, DESIGNER, SUPPLEMENTAL MATERIAL

The contents of this issue are adaptations of material previously published in SCIENTIFIC AMERICAN.

SCIENTIFIC AMERICAN EDITORIAL ADVISORS
John Rennie, EDITOR IN CHIEF
Mariette DiChristina, EXECUTIVE EDITOR
Gary Stix, SPECIAL PROJECTS EDITOR

PRODUCTION
William Sherman, ASSOCIATE PUBLISHER, PRODUCTION
Silvia Di Placido, PREPRESS AND QUALITY MANAGER
Georgina Franco, PRINT PRODUCTION MANAGER
Christina Hippeli, PRODUCTION MANAGER

ANCILLARY PRODUCTS
Diane McGarvey, DIRECTOR

CIRCULATION
Lorraine Leib Terlecki, ASSOCIATE PUBLISHER/ VICE PRESIDENT, CIRCULATION

BUSINESS ADMINISTRATION
Michael Florek, GENERAL MANAGER
Marie Maher, BUSINESS MANAGER

CHAIRMAN EMERITUS
John J. Hanley

CHAIRMAN
Rolf Grisebach

PRESIDENT AND CHIEF EXECUTIVE OFFICER
Gretchen G. Teichgraeber

VICE PRESIDENT
Frances Newburg

By Walter C. Willett
and Meir J. Stampfer

REBUILD
the Food Pyramid

THE DIETARY GUIDE
INTRODUCED A DECADE AGO
HAS LED PEOPLE ASTRAY. SOME
FATS ARE HEALTHY FOR THE HEART,
AND MANY CARBOHYDRATES CLEARLY ARE NOT

ING

In 1992

the U.S. Department of Agriculture officially released the Food Guide Pyramid, which was intended to help the American public make dietary choices that would maintain good health and reduce the risk of chronic disease. The recommendations embodied in the pyramid soon became well known: people should minimize their consumption of fats and oils but should eat six to 11 servings a day of foods rich in complex carbohydrates—bread, cereal, rice, pasta and so on. The food pyramid also recommended generous amounts of vegetables (including potatoes, another plentiful source of complex carbohydrates), fruit and dairy products, and at least two servings a day from the meat and beans group, which lumped together red meat with poultry, fish, nuts, legumes and eggs.

Even when the pyramid was being developed, though, nutritionists had long known that some types of fat are essential to health and can reduce the risk of cardiovascular disease. Furthermore, scientists had found little evidence that a high intake of carbohydrates is beneficial. Since 1992 more and more research has shown that the USDA pyramid is grossly flawed. By promoting the consumption of all complex carbohydrates and eschewing all fats and oils, the pyramid provides misleading guidance. In short, not all fats are bad for you, and by no means are all complex carbohydrates good for you. The USDA's Center for Nutrition Policy and Promo-

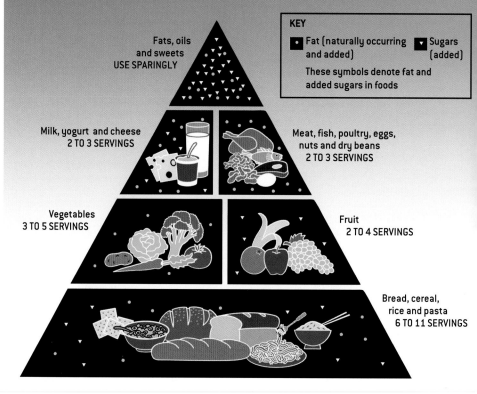

OLD FOOD PYRAMID

conceived by the U.S. Department of Agriculture was intended to convey the message "Fat is bad" and its corollary "Carbs are good." These sweeping statements are now being questioned.

For information on the amount of food that counts as one serving, visit www.nal.usda.gov:8001/py/pmap.htm

tion is now reassessing the pyramid, but this effort is not expected to be completed until 2004. In the meantime, we have drawn up a new pyramid that better reflects the current understanding of the relation between diet and health. Studies indicate that adherence to the recommendations in the revised pyramid can significantly reduce the risk of cardiovascular disease for both men and women.

How did the original USDA pyramid

go so wrong? In part, nutritionists fell victim to a desire to simplify their dietary recommendations. Researchers had known for decades that saturated fat—found in abundance in red meat and dairy products—raises cholesterol levels in the blood. High cholesterol levels, in turn, are associated with a high risk of coronary heart disease (heart attacks and other ailments caused by the blockage of the arteries to the heart). In the 1960s controlled feeding studies, in which the participants eat carefully prescribed diets for several weeks, substantiated that saturated fat increases cholesterol levels. But the studies also showed that polyunsaturated fat—found in vegetable oils and fish—reduces cholesterol. Thus, dietary advice during the 1960s and 1970s emphasized the replacement of saturated fat with polyunsaturated fat, not total fat reduction. (The subsequent doubling of polyunsaturated fat consumption among Americans probably contributed greatly to the halving of coronary heart disease rates in the U.S. during the 1970s and 1980s.)

Overview/*The Food Guide Pyramid*

- The U.S. Department of Agriculture's Food Guide Pyramid, introduced in 1992, recommended that people avoid fats but eat plenty of carbohydrate-rich foods such as bread, cereal, rice and pasta. The goal was to reduce the consumption of saturated fat, which raises cholesterol levels.
- Researchers have found that a high intake of refined carbohydrates such as white bread and white rice can wreak havoc on the body's glucose and insulin levels. Replacing these carbohydrates with healthy fats—monounsaturated or polyunsaturated—actually lowers one's risk of heart disease.
- Nutritionists are now proposing a new food pyramid that encourages the consumption of healthy fats and whole grain foods but recommends avoiding refined carbohydrates, butter and red meat.

RICHARD BORGE (*preceding pages*); USDA/DHHS (*old pyramid*)

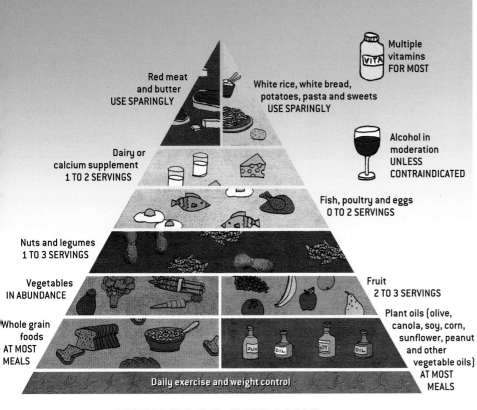

NEW FOOD PYRAMID

outlined by the authors distinguishes between healthy and unhealthy types of fat and carbohydrates. Fruits and vegetables are still recommended, but the consumption of dairy products should be limited.

The notion that fat in general is to be avoided stems mainly from observations that affluent Western countries have both high intakes of fat and high rates of coronary heart disease. This correlation, however, is limited to saturated fat. Societies in which people eat relatively large portions of monounsaturated and polyunsaturated fat tend to have lower rates of heart disease [see illustration on next page]. On the Greek island of Crete, for example, the traditional diet contained much olive oil (a rich source of monounsaturated fat) and fish (a source of polyunsaturated fat). Although fat constituted 40 percent of the calories in this diet, the rate of heart disease for those who followed it was lower than the rate for those who followed the traditional diets of Japan, in which fat made up only 8 to 10 percent of the calories. Furthermore, international comparisons can be misleading: many negative influences on health, such as smoking, physical inactivity and high amounts of body fat, are also correlated with Western affluence.

Unfortunately, many nutritionists decided it would be too difficult to educate the public about these subtleties. Instead they put out a clear, simple message: "Fat is bad." Because saturated fat represents about 40 percent of all fat consumed in the U.S., the rationale of the USDA was that advocating a low-fat diet would naturally reduce the intake of saturated fat. This recommendation was soon reinforced by the food industry, which began selling cookies, chips and other products that were low in fat but often high in sweeteners such as high-fructose corn syrup.

When the food pyramid was being developed, the typical American got about 40 percent of his or her calories from fat, about 15 percent from protein and about 45 percent from carbohydrates. Nutritionists did not want to suggest eating more protein, because many sources of protein (red meat, for example) are also heavy in saturated fat. So the "Fat is bad" mantra led to the corollary "Carbs are good." Dietary guidelines from the American Heart Association and other groups

recommended that people get at least half their calories from carbohydrates and no more than 30 percent from fat. This 30 percent limit has become so entrenched among nutritionists that even the sophisticated observer could be forgiven for thinking that many studies must show that individuals with that level of fat intake enjoyed better health than those with higher levels. But no study has demonstrated long-term health benefits that can be directly attributed to a low-fat diet. The 30 percent limit on fat was essentially drawn from thin air.

The wisdom of this direction became even more questionable after researchers found that the two main cholesterol-carrying chemicals—low-density lipoprotein (LDL), popularly known as "bad cholesterol," and high-density lipoprotein (HDL), known as "good cholesterol"—have very different effects on the risk of coronary heart disease. Increasing the ratio of LDL to HDL in the blood raises the risk, whereas decreasing the ratio lowers it. By the early 1990s controlled feeding studies had shown that when a person replaces calories from saturated fat with an equal amount of calories from carbohydrates the levels of LDL and total cholesterol fall, but the level of HDL also falls. Because the ratio of LDL to HDL does not change, there is only a small reduction in the person's risk of heart disease. Moreover, the switch to carbohydrates boosts the blood levels of triglycerides, the component molecules of fat, probably because of effects on the body's endocrine system. High triglyceride levels are also associated with a high risk of heart disease.

The effects are more grievous when a person switches from either monounsaturated or polyunsaturated fat to carbohydrates. LDL levels rise and HDL levels drop, making the cholesterol ratio worse. In contrast, replacing saturated fat with either monounsaturated or polyunsaturated fat improves this ratio and would be expected to reduce heart disease. The only fats that are significantly more deleterious than carbohydrates are the trans-unsaturated fatty acids; these are produced by the partial hydrogenation of liquid vegetable oil, which causes it to solidify.

RICHARD BORGE

Fat and Heart Disease

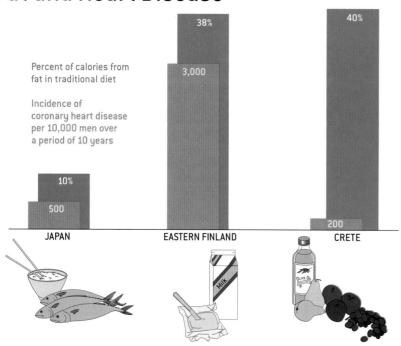

Percent of calories from fat in traditional diet

Incidence of coronary heart disease per 10,000 men over a period of 10 years

JAPAN — 10% — 500

EASTERN FINLAND — 38% — 3,000

CRETE — 40% — 200

INTERNATIONAL COMPARISONS reveal that total fat intake is a poor indicator of heart disease risk. What is important is the type of fat consumed. In regions where saturated fats traditionally made up much of the diet (for example, eastern Finland), rates of heart disease were much higher than in areas where monounsaturated fats were prevalent (such as the Greek island of Crete). Crete's Mediterranean diet, based on olive oil, was even better for the heart than the low-fat traditional diet of Japan.

Found in many margarines, baked goods and fried foods, trans fats are uniquely bad for you because they raise LDL and triglycerides while reducing HDL.

The Big Picture

TO EVALUATE FULLY the health effects of diet, though, one must look beyond cholesterol ratios and triglyceride levels. The foods we eat can cause heart disease through many other pathways, including raising blood pressure or boosting the tendency of blood to clot. And other foods can prevent heart disease in surprising ways; for instance, omega-3 fatty acids (found in fish and some plant oils) can reduce the likelihood of ventricular fibrillation, a heart rhythm disturbance that causes sudden death.

The ideal method for assessing all these adverse and beneficial effects would be to conduct large-scale trials in which individuals are randomly assigned to one diet or another and followed for many years. Because of practical constraints and cost, few such studies have been conducted, and most of these have focused on patients who already suffer from heart dis-

ease. Though limited, these studies have supported the benefits of replacing saturated fat with polyunsaturated fat, but not with carbohydrates.

The best alternative is to conduct large epidemiological studies in which the diets of many people are periodically assessed and the participants are monitored for the development of heart disease and other conditions. One of the best-known examples of this research is the Nurses' Health Study, which was begun in 1976 to evaluate the effects of oral contraceptives but was soon extended to nutrition as well. Our group at Harvard University has followed nearly 90,000 women in this study who first completed detailed questionnaires on diet in 1980, as well as more than 50,000 men who were enrolled in the Health Professionals Follow-Up Study in 1986.

After adjusting the analysis to account for smoking, physical activity and other recognized risk factors, we found that a participant's risk of heart disease was strongly influenced by the type of dietary fat consumed. Eating trans fat increased the risk substantially, and eating saturat-

ed fat increased it slightly. In contrast, eating monounsaturated and polyunsaturated fats decreased the risk—just as the controlled feeding studies predicted. Because these two effects counterbalanced each other, higher overall consumption of fat did not lead to higher rates of coronary heart disease. This finding reinforced a 1989 report by the National Academy of Sciences that concluded that total fat intake alone was not associated with heart disease risk.

But what about illnesses besides coronary heart disease? High rates of breast, colon and prostate cancers in affluent Western countries have led to the belief that the consumption of fat, particularly animal fat, may be a risk factor. But large epidemiological studies have shown little evidence that total fat consumption or intakes of specific types of fat during midlife affect the risks of breast or colon cancer. Some studies have indicated that prostate cancer and the consumption of animal fat may be associated, but reassuringly there is no suggestion that vegetable oils increase any cancer risk. Indeed, some studies have suggested that vegetable oils may slightly reduce such risks. Thus, it is reasonable to make decisions about dietary fat on the basis of its effects on cardiovascular disease, not cancer.

Finally, one must consider the impact of fat consumption on obesity, the most serious nutritional problem in the U.S. Obesity is a major risk factor for several diseases, including type 2 diabetes (also called adult-onset diabetes), coronary heart disease, and cancers of the breast, colon, kidney and esophagus. Many nutritionists believe that eating fat can contribute to weight gain because fat contains more calories per gram than protein or carbohydrates. Also, the process of storing dietary fat in the body may be more efficient than the conversion of carbohydrates to body fat. But recent controlled feeding studies have shown that these considerations are not practically important. The best way to avoid obesity is to limit your total calories, not just the fat calories. So the critical issue is whether the fat composition of a diet can influence one's ability to control caloric intake. In other words, does eating fat leave you more or less hungry than eating protein or carbohydrates? There are various theories about why one diet should be better than another, but few long-term studies have been done. In randomized trials, individuals assigned to low-fat diets tend to lose a few pounds during the first months but then regain the weight. In studies lasting a year or longer, low-fat diets have consistently not led to greater weight loss.

Carbo-Loading

NOW LET'S LOOK at the health effects of carbohydrates. Complex carbohydrates consist of long chains of sugar units such as glucose and fructose; sugars contain only one or two units. Because of concerns that sugars offer nothing but "empty calories"—that is, no vitamins, minerals or other nutrients—complex carbohydrates form the base of the USDA food pyramid. But refined carbohydrates, such as white bread and white rice, can be very quickly broken down to glucose, the primary fuel for the body. The refining process produces an easily absorbed form of starch—which is defined as glucose molecules bound together—and also removes many vitamins and minerals and fiber. Thus, these carbohydrates increase glucose levels in the blood more than whole grains do. (Whole grains have not been milled into fine flour.)

Or consider potatoes. Eating a boiled potato raises blood sugar levels higher than eating the same amount of calories from table sugar. Because potatoes are mostly starch, they can be rapidly metabolized to glucose. In contrast, table sugar (sucrose) is a disaccharide consisting of one molecule of glucose and one molecule of fructose. Fructose takes longer to convert to glucose, hence the slower rise in blood glucose levels.

A rapid increase in blood sugar stimulates a large release of insulin, the hormone that directs glucose to the muscles and liver. As a result, blood sugar plummets, sometimes even going below the baseline. High levels of glucose and insulin can have negative effects on cardiovascular health, raising triglycerides and lowering HDL (the good cholesterol). The precipitous decline in glucose can also lead to more hunger after a carbohy-

drate-rich meal and thus contribute to overeating and obesity.

In our epidemiological studies, we have found that a high intake of starch from refined grains and potatoes is associated with a high risk of type 2 diabetes and coronary heart disease. Conversely, a greater intake of fiber is related to a lower risk of these illnesses. Interestingly, though, the consumption of fiber did not lower the risk of colon cancer, as had been hypothesized earlier.

Overweight, inactive people can become resistant to insulin's effects and therefore require more of the hormone to

regulate their blood sugar. Recent evidence indicates that the adverse metabolic response to carbohydrates is substantially worse among people who already have insulin resistance. This finding may account for the ability of peasant farmers in Asia and elsewhere, who are extremely lean and active, to consume large amounts of refined carbohydrates without experiencing diabetes or heart disease, whereas the same diet in a more sedentary population can have devastating effects.

Eat Your Veggies

HIGH INTAKE OF FRUITS and vegetables is perhaps the least controversial aspect of the food pyramid. A reduction in cancer risk has been a widely promoted benefit. But most of the evidence for this benefit has come from case-control studies, in which patients with cancer and selected control subjects are asked about their earlier diets. These retrospective studies are susceptible to numerous biases, and recent findings from large prospective studies (including our own) have tended to show little relation between overall fruit and vegetable consumption and cancer incidence. (Specific nutrients in fruits and vegetables may offer benefits, though; for instance, the folic acid in green leafy vegetables may reduce the risk

of colon cancer, and the lycopene found in tomatoes may lower the risk of prostate cancer.)

The real value of eating fruits and vegetable may be in reducing the risk of cardiovascular disease. Folic acid and potassium appear to contribute to this effect, which has been seen in several epidemiological studies. Inadequate consumption of folic acid is responsible for higher risks of serious birth defects as well, and low intake of lutein, a pigment in green leafy vegetables, has been associated with greater risks of cataracts and degeneration of the retina. Fruits and vegetables are also the primary source of many vitamins needed for good health. Thus, there are good reasons to consume the recommended five servings a day, even if doing so has little impact on cancer risk. The inclusion of potatoes as a vegetable in the USDA pyra-

> The best way to avoid obesity is to
> ## LIMIT YOUR TOTAL CALORIES,
> not just the fat calories.

THE AUTHORS

WALTER C. WILLETT and *MEIR J. STAMPFER* are professors of epidemiology and nutrition at the Harvard School of Public Health. Willett chairs the school's department of nutrition, and Stampfer heads the department of epidemiology. Willett and Stampfer are also professors of medicine at Harvard Medical School. Both of them practice what they preach by eating well and exercising regularly.

Benefits of the New Pyramid

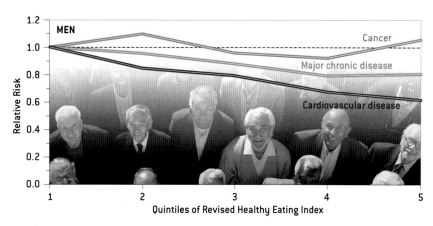

HEALTH EFFECTS OF THE RECOMMENDATIONS in the revised food pyramid were gauged by studying disease rates among 67,271 women in the Nurses' Health Study and 38,615 men in the Health Professionals Follow-Up Study. Women and men in the fifth quintile (the 20 percent whose diets were closest to the pyramid's recommendations) had significantly lower rates of cardiovascular disease than those in the first quintile (the 20 percent who strayed the most from the pyramid). The dietary recommendations had no significant effect on cancer risk, however.

epidemiological studies indicate that they lower the risk of heart disease and diabetes. Also, people who eat nuts are actually less likely to be obese; perhaps because nuts are more satisfying to the appetite, eating them seems to have the effect of significantly reducing the intake of other foods.

Yet another concern regarding the USDA pyramid is that it promotes overconsumption of dairy products, recommending the equivalent of two or three glasses of milk a day. This advice is usually justified by dairy's calcium content, which is believed to prevent osteoporosis and bone fractures. But the highest rates of fractures are found in countries with high dairy consumption, and large prospective studies have not shown a lower risk of fractures among those who eat plenty of dairy products. Calcium is an essential nutrient, but the requirements for bone health have probably been overstated. What is more, we cannot assume that high dairy consumption is safe: in several studies, men who consumed large amounts of dairy products experienced an increased risk of prostate cancer, and in some studies, women with high intakes had elevated rates of ovarian cancer. Although fat was initially assumed to be the responsible factor, this has not been supported in more detailed analyses. High calcium intake itself seemed most clearly related to the risk of prostate cancer.

More research is needed to determine the health effects of dairy products, but at the moment it seems imprudent to recommend high consumption. Most adults who are following a good overall diet can get the necessary amount of calcium by consuming the equivalent of one glass of milk a day. Under certain circumstances, such as after menopause, people may need more calcium than usual, but it can be obtained at lower cost and without saturated fat or calories by taking a supplement.

A Healthier Pyramid

ALTHOUGH THE USDA'S food pyramid has become an icon of nutrition over the past decade, until recently no studies had evaluated the health of individuals who followed its guidelines. It very likely has some benefits, especially from a high intake of fruits and vegetables. And a de-

mid has little justification, however; being mainly starch, potatoes do not confer the benefits seen for other vegetables.

Another flaw in the USDA pyramid is its failure to recognize the important health differences between red meat (beef, pork and lamb) and the other foods in the meat and beans group (poultry, fish, legumes, nuts and eggs). High consumption of red meat has been associated with an increased risk of coronary heart disease, probably because of its high content of saturated fat and cholesterol. Red meat also raises the risk of type 2 diabetes and colon cancer. The elevated risk of colon cancer may be related in part to the carcinogens produced during cooking and the chemicals found in processed meats such as salami and bologna.

Poultry and fish, in contrast, contain less saturated fat and more unsaturated fat than red meat does. Fish is a rich source of the essential omega-3 fatty acids as well. Not surprisingly, studies have shown that people who replace red meat with chicken and fish have a lower risk of coronary heart disease and colon cancer. Eggs are high in cholesterol, but consumption of up to one a day does not appear to have adverse effects on heart disease risk (except among diabetics), probably because the effects of a slightly higher cholesterol level are counterbalanced by other nutritional benefits. Many people have avoided nuts because of their high fat content, but the fat in nuts, including peanuts, is mainly unsaturated, and walnuts in particular are a good source of omega-3 fatty acids. Controlled feeding studies show that nuts improve blood cholesterol ratios, and

KEN FISHER (men) AND ED HONOWITZ (women) Getty Images; CORNELIA BLIK (graph)

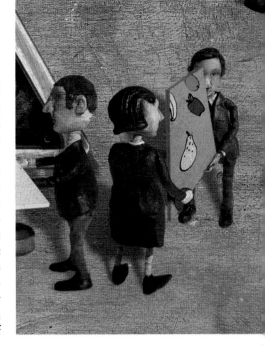

crease in total fat intake would tend to reduce the consumption of harmful saturated and trans fats. But the pyramid could also lead people to eat fewer of the healthy unsaturated fats and more refined starches, so the benefits might be negated by the harm.

To evaluate the overall impact, we used the Healthy Eating Index (HEI), a score developed by the USDA to measure adherence to the pyramid and its accompanying dietary guidelines in federal nutrition programs. From the data collected in our large epidemiological studies, we calculated each participant's HEI score and then examined the relation of these scores to subsequent risk of major chronic disease (defined as heart attack, stroke, cancer or nontraumatic death from any cause). When we compared people in the same age groups, women and men with the highest HEI scores did have a lower risk of major chronic disease. But these individuals also smoked less, exercised more and had generally healthier lifestyles than the other participants. After adjusting for these variables, we found that participants with the highest HEI scores did not experience significantly better overall health outcomes. As predicted, the pyramid's harms counterbalanced its benefits.

Because the goal of the pyramid was a worthy one—to encourage healthy dietary choices—we have tried to develop an alternative derived from the best available knowledge. Our revised pyramid [*see illustration on page 5*] emphasizes weight control through exercising daily and avoiding an excessive total intake of calories. This pyramid recommends that the bulk of one's diet should consist of healthy fats (liquid vegetable oils such as olive, canola, soy, corn, sunflower and peanut) and healthy carbohydrates (whole grain foods such as whole wheat bread, oatmeal and brown rice). If both the fats and carbohydrates in your diet are healthy, you probably do not have to worry too much about the percentages of total calories coming from each. Vegetables and fruits should also be eaten in abundance. Moderate amounts of healthy sources of protein (nuts, legumes, fish, poultry and eggs) are encouraged, but dairy consumption should be limited to one to two servings a day. The revised pyramid recommends minimizing the consumption of red meat, butter, refined grains (including white bread, white rice and white pasta), potatoes and sugar.

Trans fat does not appear at all in the pyramid, because it has no place in a healthy diet. A multiple vitamin is suggested for most people, and moderate alcohol consumption can be a worthwhile option (if not contraindicated by specific health conditions or medications). This last recommendation comes with a caveat: drinking no alcohol is clearly better than drinking too much. But more and more studies are showing the benefits of

Men and women eating in accordance with THE NEW PYRAMID had a lower risk of major chronic disease.

moderate alcohol consumption (in any form: wine, beer or spirits) to the cardiovascular system.

Can we show that our pyramid is healthier than the USDA's? We created a new Healthy Eating Index that measured how closely a person's diet followed our recommendations. Applying this revised index to our epidemiological studies, we found that men and women who were eating in accordance with the new pyramid had a lower risk of major chronic disease [*see illustration on opposite page*]. This benefit resulted almost entirely from significant reductions in the risk of cardiovascular disease—up to 30 percent for women and 40 percent for men. Following the new pyramid's guidelines did not, however, lower the risk of cancer. Weight control and physical activity, rather than specific food choices, are associated with a reduced risk of many cancers.

Of course, uncertainties still cloud our understanding of the relation between diet and health. More research is needed to examine the role of dairy products, the health effects of specific fruits and vegetables, the risks and benefits of vitamin supplements, and the long-term effects of diet during childhood and early adult life. The interaction of dietary factors with genetic predisposition should also be investigated, although its importance remains to be determined.

Another challenge will be to ensure that the information about nutrition given to the public is based strictly on scientific evidence. The USDA may not be the best government agency to develop objective nutritional guidelines, because it may be too closely linked to the agricultural industry. The food pyramid should be rebuilt in a setting that is well insulated from political and economic interests. SA

MORE TO EXPLORE

Primary Prevention of Coronary Heart Disease in Women through Diet and Lifestyle. Meir J. Stampfer, Frank B. Hu, JoAnn E. Manson, Eric B. Rimm and Walter C. Willett in *New England Journal of Medicine*, Vol. 343, No. 1, pages 16–22; July 6, 2000.

Eat, Drink, and Be Healthy: The Harvard Medical School Guide to Healthy Eating. Walter C. Willett, P. J. Skerrett and Edward L. Giovannucci. Simon & Schuster, 2001.

Dietary Reference Intakes for Energy, Carbohydrates, Fiber, Fat, Protein and Amino Acids (Macronutrients). Food and Nutrition Board, Institute of Medicine, National Academy of Sciences. National Academies Press, 2002. Available online at www.nap.edu/books/0309085373/html/

Rebuilding the Food Pyramid
IN REVIEW

TESTING YOUR COMPREHENSION

1) Which of the following is NOT a feature of the USDA's 1992 Food Guide Pyramid?
   a) It recommends that carbohydrates be eaten more often than any other food category.
   b) It recommends that fats, oils, and sweets should be eaten less often than any other food category.
   c) It recommends that dairy foods (milk, cheese, etc.) be eaten as often as high-protein foods (meat, fish, eggs, etc.).
   d) It recommends that fats from fish products be eaten more often than fats from meat products.

2) Which of the following is NOT an objection raised by nutritionists about the 1992 Food Guide Pyramid?
   a) While some fats are unhealthy, others are an essential part of the diet and have helpful effects.
   b) The recommended consumption of complex carbohydrates is too high.
   c) The recommended consumption of meat is too low.
   d) The recommended consumption of milk and milk products is too high.

3) Which of the following statements is true concerning the correlation between fat consumption and coronary heart disease?
   a) High levels of saturated fats are associated with high risk of heart disease.
   b) Low levels of saturated fats are associated with high risk of heart disease.
   c) High levels of polyunsaturated fats are associated with high risk of heart disease.
   d) Low levels of polyunsaturated fats are associated with low risk of heart disease.

4) The majority of calories in a typical American diet come from
   a) saturated fats.
   b) unsaturated fats.
   c) proteins.
   d) carbohydrates.

5) What is the most desirable relative proportion of LDL ("bad cholesterol") to HDL ("good cholesterol")?
   a) low LDL and low HDL
   b) low LDL and high HDL
   c) high LDL and low HDL
   d) high LDL and high HDL

6) Long-term epidemiological studies have revealed several important correlations between diet and health. Which of the following statements is NOT supported by such evidence?
   a) Eating trans fats increases the risk of heart disease.
   b) Eating unsaturated fats decreases the risk of heart disease.
   c) Eating saturated fats slightly increases the risk of heart disease.
   d) Eating saturated fats increases the risk of breast and colon cancer.

7) What is the best way to avoid obesity?
   a) Limit the total number of calories in the diet.
   b) Limit the number of calories from fat.
   c) Limit the number of calories from carbohydrates.
   d) Limit the number of calories from protein.

8) In what sense are simple sugars "empty calories"?
   a) The body never uses the calories.
   b) They have fewer calories than an equivalent amount of carbohydrates.
   c) They typically contain few nutrients.
   d) They are stored in the body's empty spaces.

9) According to the authors, which of the following vegetables confers the least health benefit?
   a) spinach
   b) potatoes
   c) tomatoes
   d) kale (a green, leafy vegetable)

10) According to the data presented in the graphs on page 8, which of the following statements is true?

    a) Eating according to the revised food pyramid corresponded to lower risk of cancer in men and women.

    b) Eating according to the revised food pyramid corresponded to lower risk of heart disease in men and women.

    c) Eating according to the revised food pyramid corresponded to higher risk of cancer in men and women.

    d) Eating according to the revised food pyramid corresponded to lower risk of heart disease in men but higher rates of heart disease in women.

BIOLOGY IN SOCIETY

1) Evidence that trans fats are associated with significant health risks first appeared in the 1990s. Yet, the FDA will not require food labels to contain information about the levels of trans fats until 2006. What factors do you think might have contributed to this delay? Do you think that economic considerations should affect food labeling decisions? Why or why not? How would you balance such economic factors against potential health risks or benefits?

2) In our society today, there are many different diet plans touted to help a person lose weight and be healthier. Choose a few such popular diets and compare them against the USDA's 1992 Food Guide Pyramid, and against the authors' New Food Pyramid. How does each diet hold up when compared to each set of recommendations?

3) There are many resources (books, software, websites, etc.) available to help you evaluate your own diet. Track your diet for one week and record the results. In what ways do you have a healthy diet? In what ways could your diet be improved? What changes could you make to your diet to bring it into closer agreement with the authors' New Food Pyramid?

THINKING ABOUT SCIENCE

1) It is fairly difficult to collect reliable data on the correlation between human diet and long-term health. In general, most diet-and-health studies rely on one of two methods: controlled feeding studies, in which individuals are told what to eat and in what amounts, and dietary logs, in which people record what they eat each day. What difficulties do you think each of these two methods presents in terms of scientific reliability? How might these difficulties be addressed and reduced? What would be an ideal method for testing the correlation between human diet and long-term health? Why isn't such a method practical?

2) What is the difference between a "case-control study," a "controlled feeding study," and an "epidemiological study?" What benefits do you think would be associated with each type of study? What drawbacks? What types of biases might affect each one?

WRITING ABOUT BIOLOGY

Write an essay that compares and contrasts the typical diet of individuals from two or more modern cultures. What cultural and historical factors can account for the diets found in each culture? How do the diets compare in terms of the food pyramid guidelines?

Testing Your Comprehension Answers:
1d, 2c, 3a, 4d, 5b, 6d, 7a, 8c, 9b, 10b.

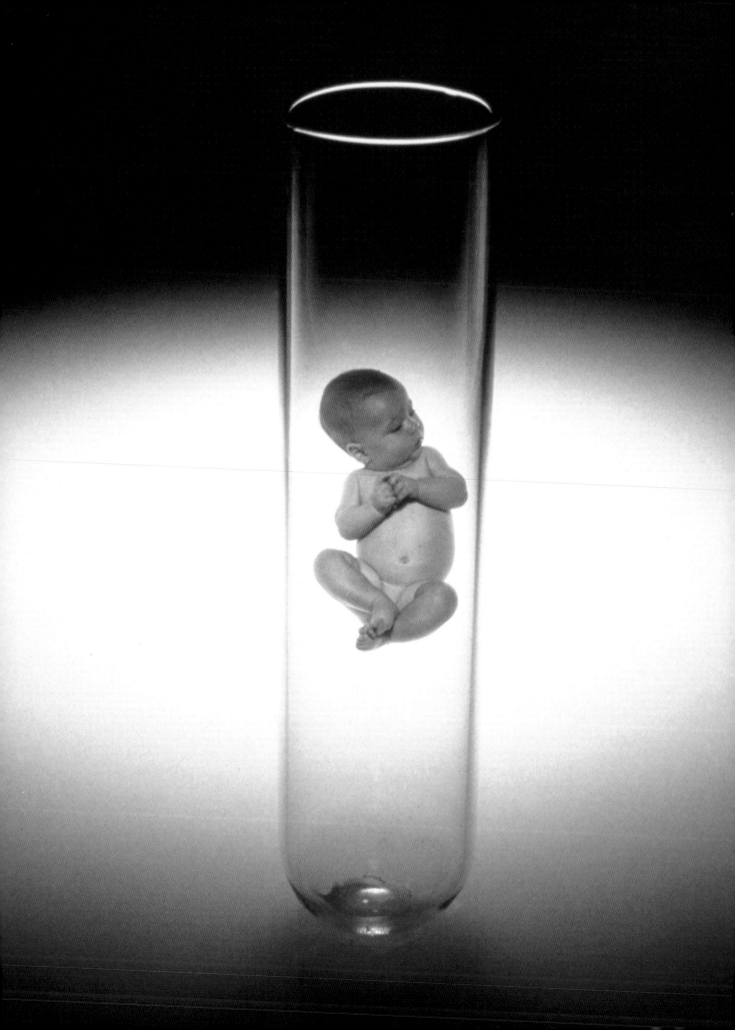

PANDORA'S BABY

In vitro fertilization was once considered by some to be a threat to our very humanity. Cloning inspires similar fears

BY ROBIN MARANTZ HENIG

On July 25, a once unique person will turn 25. This nursery school aide in the west of England seems like an average young woman, a quiet, shy blonde who enjoys an occasional round of darts at the neighborhood pub. But Louise Brown's birth was greeted by newspaper headlines calling her the "baby of the century." Brown was the world's first test tube baby.

Today people may remember Brown's name, or that she was British, or that her doctors, Steptoe and Edwards, sounded vaguely like a vaudeville act. But the past quarter of a century has dimmed the memory of one of the most important aspects of her arrival: many people were horrified by it. Even some scientists feared that Patrick Steptoe and Robert Edwards might have brewed pestilence in a petri dish. Would the child be normal, or would the laboratory manipulations leave dreadful genetic derangements? Would she be psychologically scarred by the knowledge of how bizarrely she had been created? And was she a harbinger of a race of unnatural beings who might eventually be fashioned specifically as a means to nefarious ends?

Now that in vitro fertilization (IVF) has led to the birth of an estimated one million babies worldwide, these fears and speculations may seem quaint and even absurd. But the same concerns once raised about IVF are being voiced, sometimes almost verbatim, about human cloning. Will cloning go the way of IVF, morphing from the monstrous to the mundane? And if human cloning, as well as other genetic interventions on the embryo, does someday become as commonplace as test tube baby–making, is that to be feared—or embraced? The lessons that have been

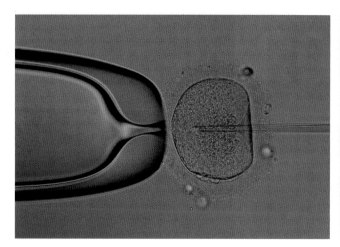

MICRONEEDLE INJECTS a sperm's package of DNA directly into a human egg, thereby achieving in vitro fertilization (*left*). The first human being born as a result of IVF, Louise Brown was 14 months old when she frolicked on the set of the *Donahue* television program (*right*). With her was Vanderbilt University IVF researcher Pierre Soupart, who predicted that "by the time Louise is 15, there will be so many others it won't be remarkable anymore."

learned from the IVF experience can illuminate the next decisions to be made.

Then and Now

AS IVF MOVED FROM the hypothetical to the actual, some considered it to be nothing more than scientists showing off: "The development of test tube babies," one critic remarked, "can be compared to the perfecting of wing transplants so that pigs might fly." But others thought of IVF as a perilous insult to nature. The British magazine *Nova* ran a cover story in the spring of 1972 suggesting that test tube babies were "the biggest threat since the atom bomb" and demanding that the public rein in the unpredictable scientists. "If today we do not accept the responsibility for directing the biologist," the *Nova* editors wrote, "tomorrow we may pay a bitter price—the loss of free choice and, with it, our humanity. We don't have much time left."

A prominent early enemy of IVF was Leon Kass, a biologist at the University of Chicago who took a professional interest in the emerging field of bioethics. If society allowed IVF to proceed, he wrote

shortly after Louise Brown's birth, some enormous issues were at stake: "the idea of the humanness of our human life and the meaning of our embodiment, our sexual being, and our relation to ancestors and descendants."

Now read Kass, a leading detractor of every new form of reproductive technology for the past 30 years, in 2003: "[Cloning] threatens the dignity of human procreation, giving one generation unprecedented genetic control over the next," he wrote in the *New York Times*. "It is the first step toward a eugenic world in which children become objects of manipulation and products of will." Such commentary coming from Kass is particularly noteworthy because of his unique position: for the past two years he has been the head of President George W. Bush's Council on Bioethics, whose first task was to offer advice on how to regulate human cloning.

Of course, IVF did not wind up creating legions of less than human children, nor did it play a role in the disintegration of the nuclear family, consequences that people like Kass feared. And so many

newer, more advanced methods of assisted reproduction have been introduced in the past decade that the "basic IVF" that produced Louise Brown now seems positively routine. One early prediction, however, did turn out to contain more than a kernel of truth. In the 1970s critics cautioned that IVF would set us tumbling down the proverbial slippery slope toward more sophisticated and, to some, objectionable forms of reproductive technology—and that once we opened the floodgates by allowing human eggs to be fertilized in the laboratory, there would be no stopping our descent.

If you consider all the techniques that might soon be available to manipulate a developing embryo, it could appear that the IVF naysayers were correct in their assessment of the slipperiness of the slope. After all, none of the genetic interventions now being debated—prenatal genetic diagnosis, gene insertions in sex cells or embryos to correct disease, the creation of new embryonic stem cell lines and, the elephant in the living room, cloning— would even be potentialities had scientists not first learned how to fertilize human eggs in a laboratory dish.

But does the existence of a such a slippery slope mean that present reproductive technology research will lead inevitably to developments that some find odious, such as embryos for tissue harvesting, or the even more abhorrent manufacture of human-nonhuman hybrids and human clones? Many people clearly fear so, which explains the current U.S. efforts to

Overview/*In Vitro Veritas*

- Many arguments against in vitro fertilization in the past and cloning today emphasize a vague threat to the very nature of humanity.
- Critics of IVF attempted to keep the federal government from supporting the research and thus ironically allowed it to flourish with little oversight.
- Because of the lack of oversight, it is only in the past few years that the increased rate of birth defects and low birth weight related to IVF have come to light.

MEMBERS of the Christian Defense Coalition and the National Clergy Council protest Advanced Cell Technologies's human cloning research outside the biotechnology firm's headquarters in Worcester, Mass., on November 30, 2001. Similar protests against IVF occurred in the 1970s.

curtail scientists' ability to manipulate embryos even before the work gets under way. But those efforts raise the question of whether science that has profound moral and ethical implications should simply never be done. Or should such science proceed, with careful attention paid to the early evolution of certain areas of research so that society can make informed decisions about whether regulation is needed?

IVF Unbound

THE FRENZY TO TRY to regulate or even outlaw cloning is in part a deliberate attempt not to let it go the way of IVF, which has been a hodgepodge of unregulated activities with no governmental or ethical oversight and no scientific coordination. Ironically, the reason IVF became so ubiquitous and uncontrolled in the U.S. was that its opponents, particularly antiabortion activists, were trying to stop it completely. Antiabortion activists' primary objection to IVF was that it involved the creation of extra embryos that would ultimately be unceremoniously destroyed—a genocide worse than at any abortion clinic, they believed. Accordingly, they thought that their best strategy would be to keep the federal government from financing IVF research.

A succession of presidential commissions starting in 1973 debated the ethics of IVF but failed to clarify matters. Some of the commissions got so bogged down in abortion politics that they never managed to hold a single meeting. Others concluded that IVF research was ethically acceptable as long as scientists honored the embryo's unique status as a "potential human life," a statement rather than a practical guideline. In 1974 the government banned federal funding for fetal research. It also forbade funding for research on the human embryo (defined as a fetus less than eight weeks old), which includes IVF. In 1993 President Bill Clinton signed the NIH Revitalization Act, which allowed federal funding of IVF research. (In 1996, however, Congress again banned embryo research.) The bottom line is that despite a series of recommendations from federal bioethics panels stating that taxpayer support of IVF research would be acceptable with certain safeguards in place, the government has nev-

er sponsored a single research grant for human IVF.

This lack of government involvement—which would also have served to direct the course of IVF research—led to a funding vacuum, into which rushed entrepreneurial scientists supported by private money. These free agents did essentially whatever they wanted and whatever the market would bear, turning IVF into a cowboy science driven by the marketplace and undertaken without guidance. The profession attempted to regulate itself—in 1986, for example, the American Fertility Society issued ethical and clinical guidelines for its members—but voluntary oversight was only sporadically effective. The quality of clinics, of which there were more than 160 by 1990, remained spotty, and those seeking IVF had little in the way of objective information to help them choose the best ones.

Today, in what appears to be an effort to avoid the mistakes made with IVF, the federal government is actively involved in regulating cloning. With the announcement in 1997 of the birth of Dolly, the first mammal cloned from an adult cell, President Clinton established mechanisms, which remain in place, to prohibit such activities in humans. Congress has made several attempts to outlaw human cloning, most recently with a bill that would make any form of human cloning punishable by a $1-million fine and up to 10 years in prison. (The House of Representatives passed this bill this past winter, but the Senate has yet to debate it.) Politicians thus lumped together two types of cloning that scientists have tried to keep separate: "therapeutic," or "research," cloning, designed to produce embryonic stem cells that might eventually mature into specialized human tissues to treat degenerative diseases; and "reproductive" cloning, undertaken specifically to bring forth a cloned human being. A second bill now

THE AUTHOR

ROBIN MARANTZ HENIG has written seven books, most recently The Monk in the Garden: The Lost and Found Genius of Gregor Mendel. Her articles have appeared in the New York Times Magazine, Civilization and Discover, among other publications. Her honors include an Alicia Patterson Foundation fellowship and a nomination for a National Book Critics Circle Award. She lives in New York City with her husband, Jeffrey R. Henig, a political science professor at Columbia University; they have two nearly grown daughters. Her next book, entitled Pandora's Baby, is about the early days of in vitro fertilization research.

WILLIAM PLOWMAN Getty

From Outrage to Approval

THE STORY of Doris Del-Zio demonstrates the ironies resulting from society's changing attitude toward IVF in the 1970s. After years of failure to conceive a child, Del-Zio and her husband turned to Landrum Shettles of what is now known as the Columbia Presbyterian Medical Center. In the fall of 1973 Shettles prepared to attempt a hasty IVF procedure on the couple. The operation was abruptly terminated by Shettles's superior, Raymond Vande Wiele, who was outraged at Shettles's audacity and who questioned the medical ethics of IVF. Vande Wiele confiscated and froze the container holding the Del-Zios' eggs and sperm. As far as the Del-Zios were concerned, Vande Wiele had committed murder: they sued him and his employers for $1.5 million.

By coincidence, the Del-Zios' case against Vande Wiele was finally brought to trial in July 1978, the same month that Louise Brown was born. The birth of the world's first test tube baby put Shettles's early IVF attempt in a different light. After Brown's appearance, most people—including the two men and four women on the Del-Zio jury—seemed much more inclined to think of IVF as a medical miracle than as a threat to civilized society.

The trial lasted six weeks, each side making its case about the wisdom, safety and propriety of IVF. In the end, Vande Wiele was found to be at fault for "arbitrary and malicious" behavior, and he and his co-defendants were ordered to pay Doris Del-Zio $50,000.

IVF developed rapidly after the trial, and 200 more test tube babies—including Louise Brown's sister, Natalie—were born over the next five years. (Natalie is now a mother, having conceived naturally, and is the first IVF baby to have a child.) Seeing so many healthy-looking test tube babies worldwide changed Vande Wiele's opinion, a change that paralleled the transformation in feeling about IVF that was occurring in the public at large. When Columbia University opened the first IVF clinic in New York City in 1983, its co-director was Raymond Vande Wiele. —R.M.H.

COURTING JUSTICE: Doris Del-Zio and her attorney, Michael Dennis, outside U.S. district court in New York City on July 17, 1978, after a session of jury selection. Del-Zio and her husband, John, sued physician Raymond Vande Wiele for derailing their early attempt at in vitro fertilization.

before the Senate would explicitly protect research cloning while making reproductive cloning a federal offense.

IVF Risks Revealed

ONE RESULT OF the unregulated nature of IVF is that it took nearly 25 years to recognize that IVF children *are* at increased medical risk. For most of the 1980s and 1990s, IVF was thought to have no effect on birth outcomes, with the exception of problems associated with multiple births: one third of all IVF pregnancies resulted in twins or triplets, the unintended consequence of the widespread practice of implanting six or eight or even 10 embryos into the womb during each IVF cycle, in the hope that at least one of them would "take." (This brute-force method also leads to the occasional set of quadruplets.) When early studies raised concerns about the safety of IVF—showing a doubling of the miscarriage rate, a tripling of the rate of stillbirths and neonatal deaths, and a fivefold increase in ectopic pregnancies—many people attributed the problems not to IVF itself but to its association with multiple pregnancies.

By last year, however, IVF's medical dark side became undeniable. In March 2002 the *New England Journal of Medicine* published two studies that controlled for the increased rate of multiple births among IVF babies and still found problems. One study compared the birth weights of more than 42,000 babies conceived through assisted reproductive technology, including IVF, in the U.S. in 1996 and 1997 with the weights of more than three million babies conceived naturally. Excluding both premature births and multiple births, the test tube babies were still two and a half times as likely to have low birth weights, defined as less than 2,500 grams, or about five and a half pounds. The other study looked at more than 5,000 babies born in Australia between 1993 and 1997, including 22 percent born as a result of IVF. It found that IVF babies were twice as likely as naturally conceived infants to have multiple major birth defects, in particular chromosomal and musculoskeletal abnormalities. The Australian researchers speculate that these problems may be a consequence

THE DAY AFTER her 20th birthday, Louise Brown poses at home with her parents.

of the drugs used to induce ovulation or to maintain pregnancy in its early stages. In addition, factors contributing to infertility may increase the risk of birth defects. The technique of IVF itself also might be to blame. A flawed sperm injected into an egg, as it is in one IVF variation, may have been unable to penetrate the egg on its own and is thus given a chance it would otherwise not have to produce a baby with a developmental abnormality.

Clearly, these risks could remain hidden during more than two decades of experience with IVF only because no system was ever put in place to track results. "If the government had supported IVF, the field would have made much more rapid progress," says Duane Alexander, director of the National Institute of Child Health and Human Development. "But as it is, the institute has never funded human IVF research of any form"—a record that Alexander calls both incredible and embarrassing.

Although the medical downsides of IVF are finally coming to light, many of the more alarmist predictions about where IVF would lead never came to pass. For example, one scenario was that it would bring us "wombs for hire," an oppressed underclass of women paid to bear the children of the infertile rich. But surrogate motherhood turned out to be expensive and emotionally complex for all parties, and it never became widespread.

Human cloning, too, might turn out to be less frightening than we currently imagine. Market forces might make reproductive cloning impractical, and scientific advancement might make it unnecessary. For example, people unable to produce eggs or sperm might ponder cloning to produce offspring. But the technology developed for cloning could make it possible to create artificial eggs or sperm containing the woman's or man's own DNA, which could then be combined with the sperm or egg of a partner. In the future, "cloning" might refer only to what is now being called therapeutic cloning, and it might eventually be truly therapeutic: a laboratory technique for making cells for the regeneration of dam-

aged organs, for example. And some observers believe that the most common use of cloning technology will ultimately not involve human cells at all: the creature most likely to be cloned may wind up being a favorite family dog or cat.

The history of IVF reveals the pitfalls facing cloning if decision making is simply avoided. But despite similarities in societal reactions to IVF and cloning, the two technologies are philosophically quite different. The goal of IVF is to enable sexual reproduction in order to produce a genetically unique human being.

Only the site of conception changes, after which events proceed much the way they normally do. Cloning disregards sexual reproduction, its goal being to mimic not the process but the already existing living entity. Perhaps the biggest difference between IVF and cloning, however, is the focus of our anxieties. In the 1970s the greatest fear related to in vitro fertilization was that it would fail, leading to sorrow, disappointment and possibly the birth of grotesquely abnormal babies. Today the greatest fear about human cloning is that it may succeed. ▪

MORE TO EXPLORE

Moving toward Clonal Man: Is This What We Want? James D. Watson in *Atlantic Monthly*, Vol. 227, No. 5, pages 50–53; May 1971.

The Frankenstein Myth Becomes a Reality: We Have the Awful Knowledge to Make Exact Copies of Human Beings. Willard Gaylin in *New York Times Magazine*, pages 12–13, 41–46; March 5, 1972.

The Embryo Sweepstakes. David Rorvik in *New York Times Magazine*, pages 17, 50–62; September 15, 1974.

Remaking Eden: How Genetic Engineering and Cloning Will Transform the American Family. Lee M. Silver. Avon Books, 1998.

The Clone Age: Adventures in the New World of Reproductive Technology. Lori B. Andrews. Henry Holt and Company, 1999.

Free to Be Me: Would-Be Cloners Pushing the Debate. Rick Weiss in *Washington Post*, page A1; May 12, 2002.

Pandora's Baby
IN REVIEW

TESTING YOUR COMPREHENSION

1) A "test tube baby" is produced through the process of
   a) in vivo fertilization.
   b) in vitro fertilization.
   c) therapeutic cloning.
   d) reproductive cloning.

2) When was the first person born as a result of an in vitro fertilization procedure?
   a) 1862
   b) 1953
   c) 1978
   d) 1992

3) Today there are about _____ people who were born as a result of an in vitro fertilization procedure.
   a) 0
   b) 100
   c) 10,000
   d) 1,000,000

4) Today there are about _____ people who were born as a result of human cloning.
   a) 0
   b) 100
   c) 10,000
   d) 1,000,000

5) For most of the last 25 years, what has been the U.S. federal policy on IVF research?
   a) All research on IVF was banned.
   b) The federal funding of IVF research was banned.
   c) The federal government funded IVF research but failed to regulate it.
   d) The federal government funded and carefully regulated IVF research.

6) The type of cloning that intends to produce a new living human is called
   a) therapeutic cloning.
   b) reproductive cloning.
   c) in vitro cloning.
   d) in vivo cloning.

7) The type of cloning that intends to produce embryonic stem cells is called
   a) therapeutic cloning.
   b) reproductive cloning.
   c) in vitro cloning.
   d) in vivo cloning.

8) What is the current U.S. federal policy on human reproductive cloning research?
   a) All research on human reproductive cloning is banned.
   b) The federal funding of human reproductive cloning research is banned.
   c) The federal government funds reproductive cloning research but fails to regulate it.
   d) The federal government funds and carefully regulates human reproductive cloning research.

9) Why do IVF procedures increase the chance of multiple births?
   a) IVF embryos are often cloned, producing multiple embryos.
   b) Women undergoing IVF have closer medical supervision than other women.
   c) More IVF-created embryos survive than naturally created embryos.
   d) IVF usually involves the implantation of six or more embryos at once.

10) Which of the following statements about the safety of IVF is most correct?

   a) No convincing data has ever been collected indicating that IVF babies have an increased health risk.

   b) From the beginning of IVF research, it was clear that IVF babies have a small but significant increased health risk.

   c) Recent evidence suggests that IVF babies do have increased risk of low birth weight.

   d) Recent studies have found that the increased risks for IVF babies include low birth weight and major birth defects.

BIOLOGY IN SOCIETY

1) If you were asked to testify before a congressional hearing on cloning research, what would be your opinion about federal funding of therapeutic cloning? Do you think the government should fund and/or oversee such research? What about for reproductive cloning? Would you testify differently about therapeutic and reproductive cloning? Why? What regulations or safeguards would you recommend be put into place?

2) Do you think that insurance companies should pay for infertile couples to have an IVF procedure? Why or why not? What about for non-infertile couples?

3) Present three arguments in favor of human reproductive cloning. Present three arguments against it. Which do you consider more convincing and why?

THINKING ABOUT SCIENCE

1) On page 16, the author describes two studies published in 2002 about the risks of IVF. For each study, rephrase the information presented in terms of the scientific method. For each of the two studies, write a hypothesis that could have served as the basis of the experiment. Next, describe the method used and the data collected. What conclusions were drawn from that data? What possible sources of error were there in the experimental design? Can you think of a way (real or hypothetical) to get around those sources of error?

2) How might the procedures of human gene therapy and human therapeutic cloning be combined to cure a child born with a hereditary immune defect? Design a series of hypothetical procedures that might be effective.

WRITING ABOUT SCIENCE

Write an essay that imagines a future society where human cloning is as widespread and accepted as IVF is today. What problems might such a society face? What benefits?

Testing Your Comprehension Answers:
1b, 2c, 3d, 4a, 5b, 6b, 7a, 8a, 9d, 10d.

WHOSE BLOOD IS IT, ANYWAY?

Blood collected from umbilical cords and placentas—
which are usually thrown away following birth—contains
stem cells that can rebuild the blood and immune
systems of people with leukemia and other cancers

By Ronald M. Kline >> Photographs by Max Aguilera-Hellweg

Kristina Romero, four months pregnant, plans to use the cord blood for her son with leukemia, Chase.

Wrinkly-faced, slippery and

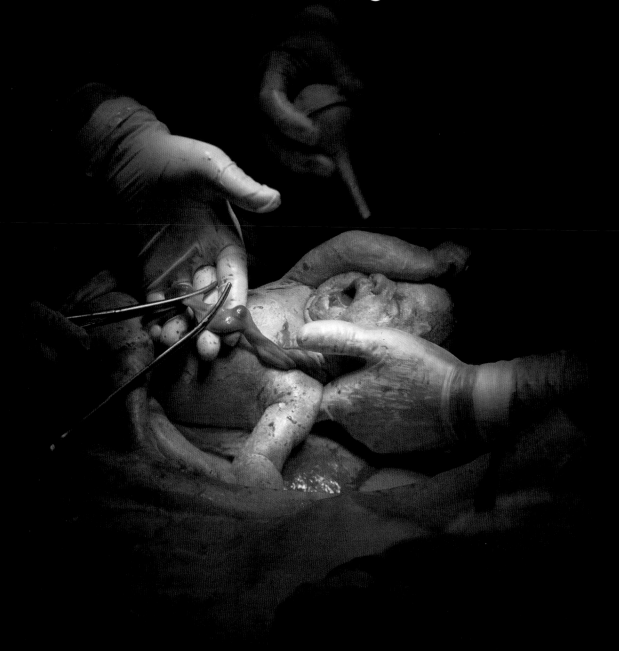

Doctors clamp the umbilical cord of a child being delivered by cesarean section.

squalling, the newborn makes her debut into the world. As the parents share their joy and begin to count 10 perfect little fingers and 10 adorable tiny toes, they scarcely pay attention to birth's Act Two: the delivery of the placenta, or afterbirth.

After the ordeal of labor, most new mothers are happy they need to push only once more for their physician to scoop up the roughly one-pound, pancakelike organ that nourished their baby through the umbilical cord for nine months. After cutting the cord and checking the afterbirth for gaps and tears that might indicate that a piece still remains inside the mother's uterus—where it could cause a potentially fatal infection—the doctor usually tosses it into a stainless-steel bucket with the rest of the medical waste bound for incineration.

But more and more physicians and parents are realizing the value of what they used to regard as merely birth's byproduct. Since 1988 hundreds of lives have been saved by the three ounces of blood contained in a typical placenta and umbilical cord. That blood is now known to be a rich source of so-called hematopoietic stem cells, the precursors of everything in the blood from infection-fighting white blood cells to the red blood cells that carry oxygen to the platelets that facilitate blood clotting after an injury.

The stem cells from a single placenta are sufficient to rebuild the blood and immune system of a child with leukemia, whose own white blood cells are abnormally dividing and must be killed by chemotherapy. In the past, physicians had to seek a living donor to provide such children with transplants of bone marrow, which also contains stem cells that produce blood and immune cells. Unfortu-

nately, many people have died during the long search for a donor with a matching tissue type or from complications if the donated marrow did not match well. Cord blood, which can be stored, is more likely to provide a suitable match and less likely to cause complications, because its stem cells are immunologically different from and more tolerant than those in adult bone marrow.

The benefits of umbilical cord blood transplantation have been demonstrated most conclusively in leukemia, but the process has other uses. The stem cells in cord blood can help to restore normal red blood cells in people with sickle cell anemia and to reconstitute the immune system of infants born with severe combined immunodeficiency. Cord blood can also be used to treat fatal inherited enzyme deficiencies, such as Hurler's syndrome, which results in progressive neurological degeneration and death. In such cases, the stem cells in cord blood can give rise not only to normal red and white blood cells but also to supporting cells in the brain called microglia that can provide the crucial missing enzyme there.

Recognizing the apparent advantages of umbilical cord blood transplantation, a number of medical centers have established banks so that a mother can donate her baby's cord blood for use by a stranger in need. The New York Blood Center's Placental Blood Program, pioneered by Pablo Rubinstein, now has 13,000 banked donations and is the nation's largest pub-

lic cord blood bank. The University of California at Los Angeles and Duke University also have umbilical cord blood storage programs, which are federally funded.

But like many new scientific discoveries, umbilical cord blood transplantation brings with it a set of ethical questions [see box on next page]. Who owns umbilical cord blood: both parents, the mother or the infant? What happens if a mother donates her baby's cord blood to a bank but the child later develops leukemia and needs it? The ethical questions are compounded by the advent of for-profit companies that collect and preserve a newborn's cord blood for possible use by the family later. Is it right for such companies to aggressively market their services—which can cost $1,500 for collection and $95 per year for storage—when the chance a child will ever need his or her cord blood ranges from 1 in 10,000 (according to the New York Blood Center) to 1 in 200,000 (according to the National Institutes of Health)?

Founts of Stem Cells

THE FIRST HINT that umbilical cord blood could be clinically useful came in 1972, when Norman Ende of the University of Medicine and Dentistry of New Jersey and his brother, Milton, a physician in Petersburg, Va., reported giving a 16-year-old leukemia patient an infusion of cord blood. Weeks later the scientists found that the patient's blood contained red cells that they could identify as hav-

PHOTOGRAPH BY MAX-AGUILERA-HELLWEG *TimePix*

But Is It Ethical?

Marketing tactics and privacy issues raise eyebrows

LAST SEPTEMBER a little girl from California named Molly received a lifesaving transplant of umbilical cord blood from her newborn brother, Adam. Molly, who was then eight years old, suffered from a potentially fatal genetic blood disorder known as Fanconi anemia. But what made the procedure particularly unusual was that Adam might not have been born had his sister not been sick. He was conceived through in vitro fertilization, and physicians specifically selected his embryo from a group of others for implantation into his mother's womb after tests showed that he would not have the disease and that he would be the best tissue match for Molly.

Was this ethically appropriate? A panel of bioethicists decided that it was, because donating cord blood would have no effect on Adam's health.

Selectively conceiving a potential donor is only one of the myriad ethical issues surrounding umbilical cord blood transplantation. One of the most significant has to do not with how the blood is used but with the marketing campaigns aimed at prospective parents by for-profit companies that offer to collect and store a baby's cord blood—for a hefty fee—in case he or she might need it later.

Such companies market cord blood collection as "biological insurance" to expectant parents. But "the odds are so extraordinarily against their child's ever needing it," says Paul Root Wolpe, a fellow at the University of Pennsylvania Health System Center for Bioethics. He fears that parents who can scarcely afford the service might feel impelled to buy it even though their families have no history of blood disorders.

Viacord, a cord blood–preserving company based in Boston, says that just five of their 6,500 clients have so far needed infusions of their stored cord blood. Moreover, only 20 percent have a family history of a blood disorder or are now in treatment.

The American Academy of Pediatrics issued a policy statement on umbilical cord blood banking in July 1999 cautioning that "it is difficult to recommend that parents store their children's cord blood for future use" unless a family member has had a blood disorder. Instead it encouraged parents to donate their baby's cord blood to public banks.

Questions have been raised in the past concerning the ownership of cord blood. But bioethicist Jeremy Sugarman of Duke University states that it is now fairly clear that although an infant owns his or her own cord blood, parents have legal guardianship over it—just as they do over the child—until he or she reaches age 18. Sugarman and Wolpe contributed to a 1997 consensus statement on the ethics of umbilical cord blood banking in the *Journal of the American Medical Association*.

Sugarman adds that it is perfectly appropriate for a parent to use one sibling's cord blood to treat another. If the first child develops a need for a transplantation later on, the fact that the parents already used his or her stored blood is unfortunate but not unethical.

Of more concern is how to ensure the safety of cord blood donated to cord banks. What happens if parents donate a newborn's cord blood to a public bank and the child develops leukemia years later? If the donated blood has no identifying information to link it to the donor, there would be no way to prevent it from being used in another child. Stem cells in the umbilical cord blood of a child who later gets leukemia could also cause leukemia in a recipient. But keeping permanent records of donors carries privacy risks: What if the blood is transplanted into a recipient but doesn't take, and the sick child's parents want to track down the donor child for bone marrow cells?

Most public cord blood banks label samples so they can be linked to a particular donor for several years, at which time they destroy the identifying information. Wolpe says that this is a good trade-off but that risks will always be associated with donor cord blood, just as they are with donor adult blood. "You try to keep it as safe as you can," he says, "but people take a chance." — *Carol Ezzell, staff writer*

ing sprung from the donor's stem cells.

But it took years for other physicians to recognize the potential of umbilical cord blood transplantation. In 1989 Hal E. Broxmeyer of the Indiana University School of Medicine, Edward A. Boyse of Memorial Sloan-Kettering Cancer Center in New York City and their colleagues revived interest in the technique by showing that human cord blood contains as many stem cells as bone marrow does. That same year Broxmeyer, Eliane Gluckman of Saint-Louis Hospital in Paris and their co-workers reported curing Fanconi anemia—a potentially fatal genetic disorder—in a five-year-old boy using blood from his baby sister's umbilical cord. Since then, approximately 75 percent of umbilical cord blood transplants have used cord blood from a nonrelative obtained from cord blood storage programs.

What's Bred in the Bone

UMBILICAL CORD BLOOD transplantation aims to obtain a source of stem cells that is the best possible match for a particular patient's tissue type. Tissue type is determined by a set of genes that make proteins called human leukocyte antigens (HLAs), which are found on the surfaces of all body cells. The immune system recognizes cells bearing the HLA proteins it has encountered since birth as normal, or belonging to the "self." Any other HLA proteins are regarded as "nonself," or foreign; cells carrying them are quickly killed.

There are six major HLA genes. Every person has two copies, or alleles, of each—one from each parent. (Each allele can come in more than 30 different types.) For bone marrow transplants, physicians aim to match the six alleles (of the total 12) that are most clinically relevant in transplantation. But because cord blood cells are immunologically different from bone marrow cells, doctors can use donor cord blood samples that match five—or even three—HLA alleles.

The genetic blueprints for making HLA proteins are found on chromosome 6. The rules of genetics dictate that the probability that two siblings will inherit the same maternal and paternal chromosome 6—and will therefore be good tissue-type matches—is only 25 percent.

Receiving a bone marrow transplant from someone who is not a good tissue-type match is potentially fatal. On one hand, the graft can fail if even a tiny amount of the recipient's own immune cells survive to generate an immune response that deems the transplanted cells foreign and kills them. This graft failure essentially leaves the patient without a many unknown minor HLA proteins. Although these proteins are not actively matched in sibling transplants either, the close genetic relationship of siblings ensures that many of them will be matched simply by chance. A good sibling pairing, however, still carries a 20 percent risk of graft-versus-host disease.

One way to slash this incidence would be to attempt to match all known HLA proteins, but that would drastically reduce the chances of finding any potential donor for a recipient. Umbilical cord blood transplantation offers a better alternative. Because of differences in the newborn's immune system, immune cells in umbilical cord blood are much less likely than those in an older child's or an adult's bone mar-

For-profit companies will preserve a newborn's cord blood for possible use by the family later. Is that right when the chance a child will ever need his or her cord blood ranges from 1 in 10,000 to 1 in 200,000?

functioning immune system and extremely vulnerable to infection. Conversely, the transplanted cells can attack the recipient's body as foreign in a dire phenomenon called graft-versus-host disease. Graft-versus-host disease can manifest itself as a blistering and ulcerating skin rash, liver damage that progresses to liver failure or severe gastrointestinal bleeding; it can quickly lead to death.

To minimize such serious complications in people who cannot obtain a bone marrow transplant from a well-matched sibling, in 1987 a coalition of national blood bank organizations persuaded the U.S. federal government to establish the National Marrow Donor Program to find the best matches for patients among a pool of registered potential bone marrow donors. The program—together with other, similar, international registries—lists 6.5 million names. But because there is only a 1 in 400 chance that an individual will be a match for someone who is not a relative, those in need typically have just a 60 percent chance of finding a potentially lifesaving donor. The odds are even worse for patients who are members of a minority group, because matches are more likely to occur between people of the same race and the registries do not have enough minority volunteers.

Even those who do find a suitable donor from one of the registries still face an alarming 80 percent risk of moderate to severe (grade II to IV) graft-versus-host disease. Scientists think this is because the matching process does not consider the

A placenta and umbilical cord ready for cord blood collection.

THE AUTHOR

RONALD M. KLINE directs the division of pediatric hematology/oncology and blood and bone marrow transplantation at Atlantic Children's Medical Center in New Jersey, where he has been since 1998. Previously he directed the umbilical cord blood transplantation program at the University of Louisville and the blood and marrow transplantation program at Kosair Children's Hospital in Louisville. He received both his undergraduate degree and his M.D. from the University of California, Los Angeles. Kline has been a vocal advocate of the use of animals in research. In 1989 he wrote an essay for *Newsweek* magazine entitled "I Am the Enemy," in which he took the animal-rights movement to task for having little compassion for human suffering.

row to attack a recipient's tissues as foreign and cause graft-versus-host disease.

In 1997 Gluckman and her colleagues found evidence that umbilical cord blood transplantation—even between an unrelated donor and recipient—is safer than bone marrow transplantation. Her group studied 143 patients who had received

lect cord blood. The New York Blood Center has been able to provide suitable donors for 85 percent of its requests using a pool of only 13,000 stored cord blood samples. The pool represents just over a single day's births in the U.S.

Cord blood also has advantages in speed. Identifying a suitable unrelated

bone marrow donor is a time-consuming process that takes an average of four months. During this period, potential donors are asked to go to donor centers to have blood drawn for tissue typing and testing for viruses such as the ones that cause AIDS and hepatitis. After a donor is selected, that individual must return, pass a physical examination, give his or her informed consent and then schedule a time for the bone marrow to be harvested from the hipbone using a needle.

In contrast, cord blood is readily available from a bank's freezer and has already undergone viral testing and tissue typing. An umbilical cord blood match can be

One day an infant born with a **genetic defect** of the bone marrow or blood may be able to have his or her umbilical **cord blood** harvested at birth, repaired by genetic engineering and then reinfused.

umbilical cord blood transplants either from relatives or from a donor program. Although the transplants ranged from fully matched to two-thirds mismatched, the incidence of life-threatening (grade III or IV) graft-versus-host disease was just 5 percent in the related group and 20 percent in the unrelated group. It caused the death of only 1 percent of the related group and 6 percent of the unrelated group. In comparison, large studies using fully matched, unrelated bone marrow donors have shown a 47 percent incidence of life-threatening graft-versus-host disease, with 70 percent of those patients (33 percent of the total) eventually dying from the disease.

Umbilical cord blood transplantation has many other potential advantages over standard bone marrow transplants. The size of the potential donor pool is much larger for cord blood than for bone marrow, for example. The National Marrow Donor Program has required more than a decade to accumulate a pool of four million individuals who have been typed for potential bone marrow donation (the other 2.5 million donors are registered in other countries). But there are four million births in the U.S. annually, each of which is a potential opportunity to col-

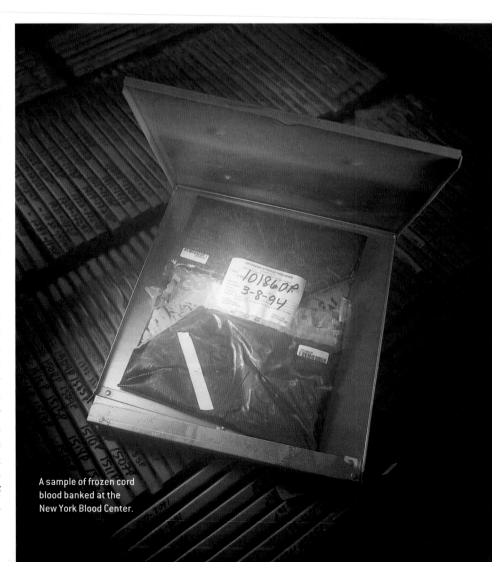

A sample of frozen cord blood banked at the New York Blood Center.

made in as few as three or four days, which can spell life or death for someone who is already immunodeficient and at high risk for a fatal infection. The collection of umbilical cord blood from as many donors as possible would also increase the likelihood that people from minority groups would be able to find a match. According to the National Marrow Donor Program, African-Americans have only a 57 percent chance of finding a bone marrow donor. Pacific Islanders and Asians have a higher match rate of 74 percent; Hispanics have a 78 percent chance; and American Indians and Alaska Natives have an 84 percent likelihood of finding a donor. Caucasians have odds of 87 percent.

Cord blood will also be virtually free of a virus that in the past has been responsible for 10 percent of deaths following bone marrow transplants: cytomegalovirus (CMV). More than half of the adult U.S. population carries CMV, which continues to live in the white blood cells of the host after initial infection. Although CMV generally causes an innocuous viral syndrome in a healthy person, it can kill someone who is immunosuppressed after a bone marrow transplant. Bone marrow donors are tested for CMV, but patients often receive CMV-positive marrow if it is the best match. Because fewer than 1 percent of infants contract CMV in the womb, umbilical cord blood could be much safer than bone marrow.

The Downside

CORD BLOOD TRANSPLANTATION is not without risks, however. One is the chance that the stem cells in a cord blood sample might harbor genetic mistakes that could cause disease in a recipient. Such disorders—which could include congenital anemias or immunodeficiencies—might not become apparent in the donor for months or years, by which time the cord blood might have already been transplanted into another recipient.

Umbilical cord blood banks could largely avoid this risk by quarantining the blood for six to 12 months and by contacting the family at that time to ensure that the donor is healthy. A long-

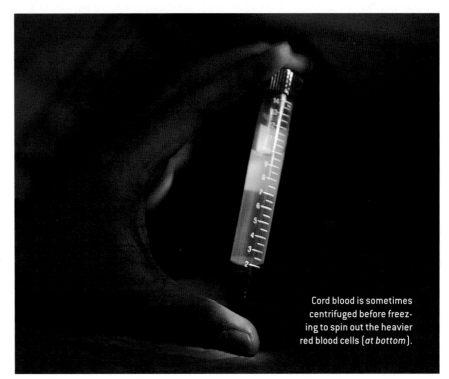

Cord blood is sometimes centrifuged before freezing to spin out the heavier red blood cells (*at bottom*).

term identification link between a donor and his or her unit of cord blood would be necessary, a prospect that has aroused privacy concerns among medical ethicists.

Currently the New York Blood Center asks potential donor parents to complete detailed questionnaires that emphasize their family histories of disease as well as their sexual histories. If responses to the questionnaire generate medical reservations, the center does not collect or store the cord blood. The center maintains only a short-term link with the donor until viral testing is complete, when the cord blood becomes anonymous.

Another limitation of umbilical cord blood is the relatively small number of stem cells contained within a single sample. Although cord blood can be used for transplantation in adults, studies by Pablo Rubinstein have demonstrated that because of the limited number of stem cells in cord blood, larger (that is, older) patients benefit less than smaller (younger) patients. Researchers are now working to devise ways to increase the number of stem cells in cord blood samples using nutrients and growth factors. They are also genetically engineering stem cells to correct genetic disorders such as severe combined immunodeficiency. In such a case, physicians would collect a patient's own cord blood, insert normal genes into the stem cells of the cord blood and reinfuse the cells into the child's body.

All of this portends even more exciting uses for cord blood. One day an infant born with a genetic defect of the bone marrow or blood may be able to have his umbilical cord blood harvested at birth, repaired by genetic engineering and then reinfused, so that he need never suffer the negative effects of his genetic inheritance. Alternatively, such a child could be cured by the infusion of stem cells from an unrelated—but perfectly matched—sample of umbilical cord blood from a donor bank. These scenarios will soon move from the realm of science fiction to science, as advances in biotechnology expand the potential of umbilical cord blood to cure diseases that once were fatal. ⓈⒶ

MORE TO EXPLORE

Ethical Issues in Umbilical Cord Blood Banking. J. Sugarman, V. Kaalund, E. Kodish, M. F. Marshall, E. G. Reisner, B. S. Wilfond and P. R. Wolpe in *Journal of the American Medical Association,* Vol. 278, No. 11, pages 938–943; September 17, 1997.

Umbilical Cord Blood Transplantation: Providing a Donor for Everyone Needing a Bone Marrow Transplant? Ronald M. Kline and Salvatore J. Bertolone in *Southern Medical Journal,* Vol. 91, No. 9, pages 821–828; September 1998.

For more details on the cord blood transplantation process, visit the University of California at Los Angeles site at **www.cordblood.med.ucla.edu**

Whose Blood is it, Anyway?
IN REVIEW

TESTING YOUR COMPREHENSION

1) Immediately after birth, blood can be collected from the umbilical cord and _____, also called the afterbirth.
   a) uterus
   b) placenta
   c) uvula
   d) stem cells

2) Hematopoietic stem cells can develop into
   a) red blood cells.
   b) white blood cells.
   c) platelets.
   d) all of the above

3) Which of these conditions is least likely to be successfully treated using cord blood stem cells?
   a) leukemia
   b) sickle cell disease
   c) breast cancer
   d) immune disorders, such as severe combined immunodeficiency (SCID)

4) The majority of umbilical cord transplants performed to date used cord blood obtained from
   a) the transplant recipient.
   b) a sibling.
   c) a non-sibling relative.
   d) a non-relative.

5) "Self" versus "non-self" cells are recognized by cell surface proteins called HLAs. HLA stands for human leukocyte
   a) antigens.
   b) antibodies.
   c) anemia.
   d) autoimmunity.

6) In terms of the HLA genes, what is the primary advantage of using umbilical cord blood for transplants compared to using bone marrow?
   a) Donor cord blood requires fewer HLA genes to match the recipient than bone marrow.
   b) Donor cord blood requires more HLA genes to match the recipient than bone marrow.
   c) Umbilical cord blood cells have more HLA genes than bone marrow cells.
   d) Umbilical cord blood cells have fewer HLA genes than bone marrow cells.

7) Assume that a mother and father are both heterozygous with different alleles for a particular HLA gene (and thus have four different alleles for that gene). What are the odds that their two children will have HLA alleles for this gene that match each other?
   a) 100%
   b) 50%
   c) 25%
   d) 0%

8) Why are immune cells extracted from umbilical cord blood better for transplantation than those in child or adult bone marrow?
   a) They are less likely to cause cancer.
   b) They are less likely to cause graft-versus-host disease.
   c) They are denser and thus easier to transplant.
   d) They can develop into more types of cells.

9) Which of the following is NOT an advantage of cord blood transplantation over bone marrow transplantation?
   a) There are many more potential cord blood donors than bone marrow donors.
   b) Finding a suitable unrelated donor can proceed more quickly for cord blood than for bone marrow.

c) Cord blood provides a much larger quantity of stem cells than bone marrow.

d) Cord blood is less likely to be infected with cytomegalovirus (CMV).

10) Of all ethnic groups, why are Caucasians the most likely to find a compatible bone marrow donor?

a) They have the most common HLA alleles.

b) They have fewer HLA alleles than other groups.

c) More Caucasians submit samples to bone marrow registries than other groups.

d) They are the least likely to suffer from graft-versus-host disease.

BIOLOGY IN SOCIETY

1) A couple gives birth to their second child, who turns out to have a life-threatening blood disease. Do you think it is ethical for the parents to give umbilical cord blood collected at their first child's birth to their second child? If their first child is not a matching donor, do you think it is ethical for the parents to have a third child (that they were otherwise not planning to have) in hopes of providing a compatible donor for their second child? If this third child were conceived using in vitro fertilization, do you think it is ethical for the parents to produce extra embryos, hoping to find one that is both disease-free and a good donor for their second child? Defend your answers to each of these questions.

2) Expectant mothers frequently receive pamphlets from private companies that will, for a substantial fee, arrange for umbilical cord blood to be collected at her child's birth and stored. These pamphlets usually include a long list of disorders that could potentially be treated with the stored stem cells. However, few of these disorders have ever ac-tually been successfully treated in the way described. Do you think it is acceptable for companies to sell this service based on potential treatments? Would you consider paying a few thousand dollars (plus a few hundred per year) for your own child's cord blood storage? What information would you wish to collect before deciding whether or not to bank your child's cord blood?

THINKING ABOUT SCIENCE

1) At the end of page 23, the author describes an early experiment involving the transplant of cord blood. In this study, researchers were able to show that the transplant recipient had blood cells that were from the donor's stem cells, rather than from the recipient's own stem cells. How could such a claim be supported? The experiment described took place in 1972. How would it be easier to support this claim today than it was 30 years ago?

2) As the Human Genome Project progresses and the functions of more genes are elucidated, many more genes that contribute to tissue rejection will be identified. How would knowing the identities of more HLA-related genes improve stem cell transplantation? How would it make it more difficult?

WRITING ABOUT SCIENCE

Write a multi-part essay that: (1) presents a hypothetical case with the clearest possible ethical argument in favor of umbilical cord blood banking; (2) presents a hypothetical case with the clearest possible ethical argument against cord blood banking; (3) describes specific issues that can be used to help determine if a particular instance of cord blood banking is ethically acceptable.

Testing Your Comprehension Answers:
1b, 2d, 3c, 4d, 5a, 6a, 7c, 8b, 9c, 10c.

On the Termination of Species

Ecologists' warnings of an ongoing mass extinction are being challenged by skeptics and largely ignored by politicians. In part that is because it is surprisingly hard to know the dimensions of the die-off, why it matters and how it can best be stopped

By W. Wayt Gibbs

CHERYL D. KNOTT

END OF AN ORANGUTAN fixes our attention and seems to confirm our worst fears about the decline of biodiversity. But does our focus on charismatic animals blur a view of the big picture? The ape in this photograph died of natural causes. And a much greater part of the earth's evolutionary heritage rises from the banks and sits in the water than lies on the log.

HILO, HAWAII—Among the scientists gathered here in August at the annual meeting of the Society for Conservation Biology, the despair was almost palpable. "I'm just glad I'm retiring soon and won't be around to see everything disappear," said P. Dee Boersma,

former president of the society, during the opening night's dinner. Other veteran field biologists around the table murmured in sullen agreement.

At the next morning's keynote address, Robert M. May, a University of Oxford zoologist who presides over the Royal Society and until last year served as chief scientific adviser to the British government, did his best to disabuse any remaining optimists of their rosy outlook. According to his latest rough estimate, the extinction rate—the pace at which species vanish—accelerated during the past 100 years to roughly 1,000 times what it was before humans showed up. Various lines of argument, he explained, "suggest a speeding up by a further factor of 10 over the next century or so.... And that puts us squarely on the breaking edge of the sixth great wave of extinction in the history of life on Earth."

From there, May's lecture grew more depressing. Biologists

and conservationists alike, he complained, are afflicted with a "total vertebrate chauvinism." Their bias toward mammals, birds and fish—when most of the diversity of life lies elsewhere—undermines scientists' ability to predict reliably the scope and consequences of biodiversity loss. It also raises troubling questions about the high-priority "hotspots" that environmental groups are scrambling to identify and preserve.

"Ultimately we have to ask ourselves why we care" about the planet's portfolio of species and its diminishment, May said. "This central question is a political and social question of values, one in which the voice of conservation scientists has no particular standing." Unfortunately, he concluded, of "the three kinds of argument we use to try to persuade politicians that all this is important . . . none is totally compelling."

Although May paints a truly dreadful picture, his is a common view for a field in which best-sellers carry titles such as *Requiem for Nature.* But is despair justified? *The Skeptical Environmentalist,* the new English translation of a recent book by Danish statistician Bjørn Lomborg, charges that reports of the death of biodiversity have been greatly exaggerated. In the face of such external skepticism, internal uncertainty and public apathy, some scientists are questioning the conservation movement's overriding emphasis on preserving rare species and the threatened hotspots in which they are concentrated. Perhaps, they suggest, we should focus instead on saving something equally at risk but even more valuable: evolution itself.

Doom ...

MAY'S CLAIM that humans appear to be causing a cataclysm of extinctions more severe than any since the one that erased the dinosaurs 65 million years ago may shock those who haven't followed the biodiversity issue. But it prompted no gasps from the conservation biologists. They have heard variations of this dire forecast since at least 1979, when Norman Myers guessed in *The Sinking Ark* that 40,000 species lose their last member each year and that one million would be extinct by 2000. In the 1980s Thomas Lovejoy similarly predicted that 15 to 20 percent would die off by 2000; Paul Ehrlich figured

Overview/*Extinction Rates*

- Eminent ecologists warn that humans are causing a mass extinction event of a severity not seen since the age of dinosaurs came to an end 65 million years ago. But paleontologists and statisticians have called such comparisons into doubt.
- It is hard to know how fast species are disappearing. Models based on the speed of tropical deforestation or on the growth of endangered species lists predict rising extinction rates. But biologists' bias toward plants and vertebrates, which represent a minority of life, undermine these predictions. Because 90 percent of species do not yet have names, let alone censuses, they are impossible to verify.
- In the face of uncertainty about the decline of biodiversity and its economic value, scientists are debating whether rare species should be the focus of conservation. Perhaps, some suggest, we should first try to save relatively pristine—and inexpensive—land where evolution can progress unaffected by human activity.

Mass Extinctions Past—and Present?

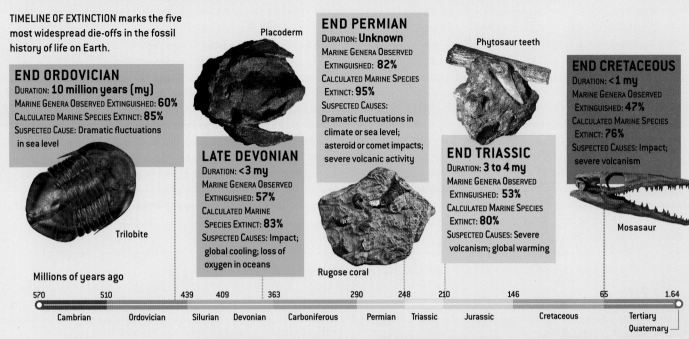

TIMELINE OF EXTINCTION marks the five most widespread die-offs in the fossil history of life on Earth.

END ORDOVICIAN
DURATION: **10 million years (my)**
MARINE GENERA OBSERVED EXTINGUISHED: **60%**
CALCULATED MARINE SPECIES EXTINCT: **85%**
SUSPECTED CAUSE: Dramatic fluctuations in sea level

Placoderm

END PERMIAN
DURATION: **Unknown**
MARINE GENERA OBSERVED
EXTINGUISHED: **82%**
CALCULATED MARINE SPECIES
EXTINCT: **95%**
SUSPECTED CAUSES:
Dramatic fluctuations in climate or sea level; asteroid or comet impacts; severe volcanic activity

Phytosaur teeth

END CRETACEOUS
DURATION: **<1 my**
MARINE GENERA OBSERVED
EXTINGUISHED: **47%**
CALCULATED MARINE SPECIES
EXTINCT: **76%**
SUSPECTED CAUSES: Impact; severe volcanism

LATE DEVONIAN
DURATION: **<3 my**
MARINE GENERA OBSERVED
EXTINGUISHED: **57%**
CALCULATED MARINE
SPECIES EXTINCT: **83%**
SUSPECTED CAUSES: Impact; global cooling; loss of oxygen in oceans

END TRIASSIC
DURATION: **3 to 4 my**
MARINE GENERA OBSERVED
EXTINGUISHED: **53%**
CALCULATED MARINE SPECIES
EXTINCT: **80%**
SUSPECTED CAUSES: Severe volcanism; global warming

Trilobite

Rugose coral

Mosasaur

Millions of years ago

| 570 | 510 | 439 | 409 | 363 | 290 | 248 | 210 | 146 | 65 | 1.64 |

| Cambrian | Ordovician | Silurian | Devonian | Carboniferous | Permian | Triassic | Jurassic | Cretaceous | Tertiary Quaternary |

With more than 1,100 species (*eight at right*) suspected to have disappeared in the past 500 years, ecologists fear a sixth mass extinction event is imminent. The die-offs so far, however, would probably not signal anything unusual to future paleontologists looking back at our time.

SPECIES *(Scientific name)*	LAST SEEN, LOCATION	EXTINCTION CAUSES
Deepwater ciscoe *(Coregonus johannae)*	1952, Lakes Huron and Michigan	Overfishing, hybridization
Pupfish *(Cyprinodon ceciliae)*	1988, Ojo de Agua La Presa, Mexico	Loss of food supply
Dobson's fruit bat *(Dobsonia chapmani)*	1970s, Cebu Islands, Philippines	Forest destruction, overhunting
Caribbean monk seal *(Monachus tropicalis)*	1950s, Caribbean Sea	Overhunting, harassment
Guam flycatcher *(Myiagra freycinetl)*	1983, Guam	Predation by introduced brown tree snakes
Kaua'i 'O'o *(Moho braccatus)*	1987, Island of Kaua'i, Hawaii	Disease, rat predation
Xerces Blue Butterfly *(Glaucopsyche xerces)*	1941, San Francisco Peninsula	Land conversion
Tobias' Caddis Fly *(Hydropsyche tobiasi)*	1950s, Rhine River, Germany	Industrial and urban pollution

SOURCES: Committee on Recently Extinct Organisms; BirdLife International; Xerces Society; World Wildlife Fund

LESTER V. BERGMAN Corbis (trilobite); JAMES L. AMOS Corbis (Placoderm); RICHARD PASELK Humboldt State University/Natural History Museum (Rugose coral and Phytosaur teeth); MIKE EVERHART Oceans of Kansas Paleontology (Mosasaur)

half would be gone by now. "I'm reasonably certain that [the elimination of one fifth of species] didn't happen," says Kirk O. Winemiller, a fish biologist at Texas A&M University who just finished a review of the scientific literature on extinction rates.

More recent projections factor in a slightly slower demise because some doomed species have hung on longer than anticipated. Indeed, a few have even returned from the grave. "It was discovered only this summer that the Bavarian vole, continental Eurasia's one and only presumed extinct mammal [since 1500], is in fact still with us," says Ross D. E. MacPhee, curator of mammalogy at the American Museum of Natural History (AMNH) in New York City.

Still, in the 1999 edition of his often-quoted book *The Diversity of Life*, Harvard University biologist E. O. Wilson cites current estimates that between 1 and 10 percent of species are extinguished every decade, at least 27,000 a year. Michael J. Novacek, AMNH's provost of science, wrote in a review article this spring that "figures approaching 30 percent extermination of all species by the mid-21st century are not unrealistic." And in a 1998 survey of biologists, 70 percent said they believed

that a mass extinction is in progress; a third of them expected to lose 20 to 50 percent of the world's species within 30 years.

"Although these assertions of massive extinctions of species have been repeated everywhere you look, they do not equate with the available evidence," Lomborg argues in *The Skeptical Environmentalist*. A professor of statistics and political science at the University of Århus, he alleges that environmentalists have ignored recent evidence that tropical deforestation is not taking the toll that was feared. "No well-investigated group of animals shows a pattern of loss that is consistent with greatly heightened extinction rates," MacPhee concurs. The best models, Lomborg suggests, project an extinction rate of 0.15 percent of species per decade, "not a catastrophe but a problem— one of many that mankind still needs to solve."

... or Gloom?

"IT'S A TOUGH question to put numbers on," Wilson allows. May agrees but says "that isn't an argument for not asking the question" of whether a mass extinction event is upon us.

To answer that question, we need to know three things: the

The Portfolio of Life

How severe is the extinction crisis? That depends in large part on how many species there are altogether. The greater the number, the more species will die out every year from natural causes and the more new ones will naturally appear. But although the general outlines of the tree of life are clear, scientists are unsure how many twigs lie at the end of each branch. When it comes to bacteria, viruses, protists and archaea (a whole kingdom of single-celled life-forms discovered just a few decades ago), microbiologists have only vague notions of how many branches there are.

Birds, fish, mammals and plants are the exceptions. Sizing up the global workforce of about 5,000 professional taxonomists, zoologist Robert M. May of the University of Oxford noted that about equal numbers study vertebrates, plants and invertebrates. "You may wish to think this record reflects some judicious appreciation of what's important," he says. "My view of that is: absolute garbage. Whether you are interested in how ecosystems evolved, their current functioning or how they are likely to respond to climate change,

you're going to learn a lot more by looking at soil microorganisms than at charismatic vertebrates."

For every group except birds, says Peter Hammond of the National History Museum in London, new species are now being discovered faster than ever before, thanks to several new international projects. An All Taxa Biodiversity Inventory under way in Great Smoky Mountains National Park in North Carolina and Tennessee has discovered 115 species—80 percent of them insects or arachnids—in its first 18 months of work. Last year 40 scientists formed the All Species Project, a society devoted to the (probably quixotic) goal of cataloguing every living species, microbes included, within 25 years.

Other projects, such as the Global Biodiversity Information Facility and Species2000, are building Internet databases that will codify species records that are now scattered among the world's museums and universities. If biodiversity is defined in strictly pragmatic terms as the variety of life-forms we know about, it is growing prodigiously.

PYRAMID OF DIVERSITY

TO A FIRST APPROXIMATION, all multicellular species are insects. Biologists know the least about the true diversity and ecological importance of the very groups that are most common.

SOURCES: Encyclopedia of Biodiversity, edited by S. A. Levin; "Biodiversity Hotspots for Conservation Priorities," by N. Myers et al. in Nature, Vol. 403, pages 853–858, February 24, 2000; William Eschemeyer (fish species); Marc Van Regenmortel (virus species); IUCN Red List 2000

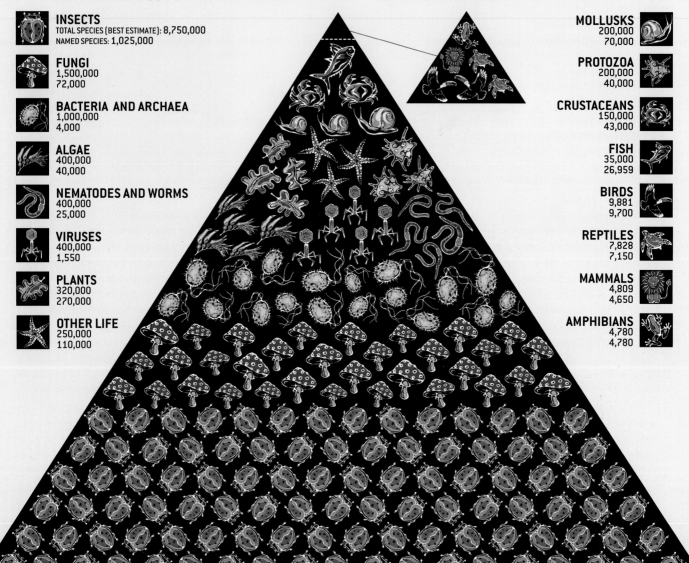

INSECTS
TOTAL SPECIES (BEST ESTIMATE): 8,750,000
NAMED SPECIES: 1,025,000

FUNGI
1,500,000
72,000

BACTERIA AND ARCHAEA
1,000,000
4,000

ALGAE
400,000
40,000

NEMATODES AND WORMS
400,000
25,000

VIRUSES
400,000
1,550

PLANTS
320,000
270,000

OTHER LIFE
250,000
110,000

MOLLUSKS
200,000
70,000

PROTOZOA
200,000
40,000

CRUSTACEANS
150,000
43,000

FISH
35,000
26,959

BIRDS
9,881
9,700

REPTILES
7,828
7,150

MAMMALS
4,809
4,650

AMPHIBIANS
4,780
4,780

natural (or "background") extinction rate, the current rate and whether the pace of extinction is steady or changing. The first step, Wilson explains, is to work out the mean life span of a species from the fossil record. "The background extinction rate is then the inverse of that. If species are born at random and all live exactly one million years—and it varies, but it's on that order—then that means one species in a million naturally goes extinct each year," he says.

In a 1995 article that is still cited in almost every scientific paper on this subject (even in Lomborg's book), May used a similar method to compute the background rate. He relied on estimates that put the mean species life span at five million to 10 million years, however; he thus came up with a rate that is five to 10 times lower than Wilson's. But according to paleontologist David M. Raup (then at the University of Chicago), who published some of the figures May and Wilson relied on, their calculations are seriously flawed by three false assumptions.

One is that species of plants, mammals, insects, marine invertebrates and other groups all exist for about the same time. In fact, the typical survival time appears to vary among groups

says, "that the current extinction rate will be sustained over millions of years." Alroy recently came up with a way to measure the speed of extinctions that doesn't suffer from such assumptions. Over the past 200 years, he figures, the rate of loss among mammal species has been some 120 times higher than natural.

A Grim Guessing Game

ATTEMPTS TO FIGURE out the current extinction rate are fraught with even more uncertainties. The international conservation organization IUCN keeps "Red Lists" of organisms suspected to be extinct in the wild. But MacPhee complains that "the IUCN methodology for recognizing extinction is not sufficiently rigorous to be reliable." He and other extinction experts have formed the Committee on Recently Extinct Organisms, which combed the Red Lists to identify those species that were clearly unique and that had not been found despite a reasonable search. They certified 60 of the 87 mammals listed by IUCN as extinct but claim that only 33 of the 92 freshwater fish presumed extinct by IUCN are definitely gone forever.

> "If you are looking for hard evidence of tens or hundreds or thousands of species disappearing each year, you aren't going to find it." —KIRK O. WINEMILLER, TEXAS A&M

by a factor of 10 or more, with mammal species among the least durable. Second, they assume that all organisms have an equal chance of making it into the fossil record. But paleontologists estimate that fewer than 4 percent of all species that ever lived are preserved as fossils. "And the species we do see are the widespread, very successful ones," Raup says. "The weak species confined to some hilltop or island all went extinct before they could be fossilized," adds John Alroy of the University of California at Santa Barbara.

The third problem is that May and Wilson use an average life span when they should use a median. Because "the vast majority of species are short-lived," Raup says, "the average is distorted by the very few that have very long life spans." All three oversimplifications underestimate the background rate—and make the current picture scarier in comparison.

Earlier this year U.C.S.B. biomathematician Helen M. Regan and several of her colleagues published the first attempt ever to correct for the strong biases and uncertainties in the data. They looked exclusively at mammals, the best-studied group. They estimated how many of the mammals now living, and how many of those recently extinguished, would show up as fossils. They also factored in the uncertainty for each number rather than relying on best guesses. In the end they concluded that "the current rate of mammalian extinction lies between 17 and 377 times the background extinction rate." The best estimate, they wrote, is a 36- to 78-fold increase.

Regan's method is still imperfect. Comparing the past 400 years with the previous 65 million unavoidably assumes, she

For every species falsely presumed absent, however, there may be hundreds or thousands that vanish unknown to science. "We are uncertain to a factor of 10 about how many species we share the planet with," May points out. "My guess would be roughly seven million, but credible guesses range from five to 15 million," excluding microorganisms.

Taxonomists have named approximately 1.8 million species, but biologists know almost nothing about most of them, especially the insects, nematodes and crustaceans that dominate the animal kingdom. Some 40 percent of the 400,000 known beetle species have each been recorded at just one location—and with no idea of individual species' range, scientists have no way to confirm its extinction. Even invertebrates known to be extinct often go unrecorded: when the passenger pigeon was eliminated in 1914, it took two species of parasitic lice with it. They still do not appear on IUCN's list.

"It is extremely difficult to observe an extinction; it's like seeing an airplane crash," Wilson says. Not that scientists aren't trying. Articles on the "biotic holocaust," as Myers calls it, usually figure that the vast majority of extinctions have been in the tropical Americas. Freshwater fishes are especially vulnerable, with more than a quarter listed as threatened. "I work in Venezuela, which has substantially more freshwater fishes than all of North America. After 30 years of work, we've done a reasonable job of cataloguing fish diversity there," observes Winemiller of Texas A&M, "yet we can't point to one documented case of extinction."

A similar pattern emerges for other groups of organisms, he

claims. "If you are looking for hard evidence of tens or hundreds or thousands of species disappearing each year, you aren't going to find it. That could be because the database is woefully inadequate," he acknowledges. "But one shouldn't dismiss the possibility that it's not going to be the disaster everyone fears."

The Logic of Loss

THE DISASTER SCENARIOS are based on several independent lines of evidence that seem to point to fast and rising extinction rates. The most widely accepted is the species-area relation. "Generally speaking, as the area of habitat falls, the number of species living in it drops proportionally by the third root to the sixth root," explains Wilson, who first deduced this equation more than 30 years ago. "A middle value is the fourth root, which means that when you eliminate 90 percent of the habitat, the number of species falls by half."

"From that rough first estimate and the rate of the destruction of the tropical forest, which is about 1 percent a year," Wilson continues, "we can predict that about one quarter of 1 percent of species either become extinct immediately or are doomed to much earlier extinction." From a pool of roughly 10 million species, we should thus expect about 25,000 to evaporate annually.

Lomborg challenges that view on three grounds, however. Species-area relations were worked out by comparing the number of species on islands and do not necessarily apply to fragmented habitats on the mainland. "More than half of Costa Rica's native bird species occur in largely deforested countryside habitats, together with similar fractions of mammals and butterflies," Stanford University biologist Gretchen Daily noted recently in Nature. Although they may not thrive, a large fraction of forest species may survive on farmland and in woodlots—for how long, no one yet knows.

That would help explain Lomborg's second observation, which is that in both the eastern U.S. and Puerto Rico, clearance of more than 98 percent of the primary forests did not wipe out half of the bird species in them. Four centuries of logging "resulted in the extinction of only one forest bird" out of 200 in the U.S. and seven out of 60 native species in Puerto Rico, he asserts.

Such criticisms misunderstand the species-area theory, according to Stuart L. Pimm of Columbia University. "Habitat destruction acts like a cookie cutter stamping out poorly mixed dough," he wrote last year in Nature. "Species found only within the stamped-out area are themselves stamped out. Those found more widely are not."

Of the 200 bird types in the forests of the eastern U.S., Pimm states, all but 28 also lived elsewhere. Moreover, the forest was cleared gradually, and gradually it regrew as farmland was abandoned. So even at the low point, around 1872, woodland covered half the extent of the original forest. The species-area theory predicts that a 50 percent reduction should knock out 16 percent of the endemic species: in this case, four birds. And four species did go extinct. Lomborg discounts one of those four that may have been a subspecies and two others that perhaps succumbed to unrelated insults.

But even if the species-area equation holds, Lomborg responds, official statistics suggest that deforestation has been slowing and is now well below 1 percent a year. The U.N. Food and Agriculture Organization recently estimated that from 1990 to 2000 the world's forest cover dropped at an average annual rate of 0.2 percent (11.5 million hectares felled, minus 2.5 million hectares of new growth).

Annual forest loss was around half a percent in most of the tropics, however, and that is where the great majority of rare and threatened species live. So although "forecasters may get these figures wrong now and then, perhaps colored by a desire to sound the alarm, this is just a matter of timescale," replies Carlos A. Peres, a Brazilian ecologist at the University of East Anglia in England.

Extinction *Filters*

SURVIVAL OF THE FITTEST takes on a new meaning when humans develop a region. Among four Mediterranean climate regions, those developed more recently have lost larger fractions of their vascular plant species in modern times. Once the species least compatible with agriculture are filtered out by "artificial selection," extinction rates seem to fall.

REGION (in order of development)	EXTINCT (per 1,000)	THREATENED (percent)
Mediterranean	1.3	14.7
South African Cape	3.0	15.2
California	4.0	10.2
Western Australia	6.6	17.5

SOURCE: "Extinctions in Mediterranean Areas." Werner Greuter in Extinction Rates. Edited by J. H. Lawton and R. H. May. Oxford University Press, 1995

An Uncertain Future

ECOLOGISTS HAVE TRIED other means to project future extinction rates. May and his co-workers watched how vertebrate species moved through the threat categories in IUCN's database over a four-year period (two years for plants), projected those very small numbers far into the future and concluded that extinction rates will rise 12- to 55-fold over the next 300 years. Georgina M. Mace, director of science at the Zoological Society of London, came to a similar conclusion by combining models that plot survival odds for a few very well known species. Entomologist Nigel E. Stork of the Natural History Museum in London noted that a British bird is 10 times more likely than a British bug to be endangered. He then extrapolated such ratios to the rest of the world to predict 100,000 to 500,000 insect extinctions by 2300. Lomborg favors this latter model, from which he concludes that "the rate for all animals will remain below 0.208 percent per decade and probably be below 0.7 percent per 50 years."

It takes a heroic act of courage for any scientist to erect such

FRANS LANTING *Minden Pictures*

long and broad projections on such a thin and lopsided base of data. Especially when, according to May, the data on endangered species "may tell us more about the vagaries of sampling efforts, of taxonomists' interests and of data entry than about the real changes in species' status."

Biologists have some good theoretical reasons to fear that even if mass extinction hasn't begun yet, collapse is imminent. At the conference in Hilo, Kevin Higgins of the University of Oregon presented a computer model that tracks artificial organisms in a population, simulating their genetic mutation rates, reproductive behavior and ecological interactions. He found that "in small populations, mutations tend to be mild enough that natural selection doesn't filter them out. That dramatically shortens the time to extinction." So as habitats shrink and populations are wiped out—at a rate of perhaps 16 million a year, Daily has estimated—"this could be a time bomb, an extinction event occurring under the surface," Higgins warns. But proving that that bomb is ticking in the wild will not be easy.

And what will happen to fig trees, the most widespread plant genus in the tropics, if it loses the single parasitic wasp variety that pollinates every one of its 900 species? Or to the 79 percent of canopy-level trees in the Samoan rain forests if hunters kill off the flying foxes on which they depend? Part of the reason so many conservationists are so fearful is that they expect the arches of entire ecosystems to fall once a few "keystone" species are removed.

WEALTH OF RAIN FORESTS, this one in Borneo, is largely unmeasured, both in biological and economic terms.

Others distrust that metaphor. "Several recent studies seem to show that there is some redundancy in ecosystems," says Melodie A. McGeoch of the University of Pretoria in South Africa, although she cautions that what is redundant today may not be redundant tomorrow. "It really doesn't make sense to think the majority of species would go down with marginally higher pressures than if humans weren't on the scene," MacPhee adds. "Evolution should make them resilient."

If natural selection doesn't do so, artificial selection might, according to work by Werner Greuter of the Free University of Berlin, Thomas M. Brooks of Conservation International and others. Greuter compared the rate of recent plant extinctions in four ecologically similar regions and discovered that the longest-settled, most disturbed area—the Mediterranean—had the lowest rate. Plant extinction rates were higher in California and South Africa, and they were highest in Western Australia. The solution to this apparent paradox, they propose, is that species that cannot coexist with human land use tend to die out soon after agriculture begins. Those that are left are better equipped to dodge the darts we throw at them. Human-induced extinctions may thus fall over time.

If true, that has several implications. Millennia ago our ancestors may have killed off many more species than we care to

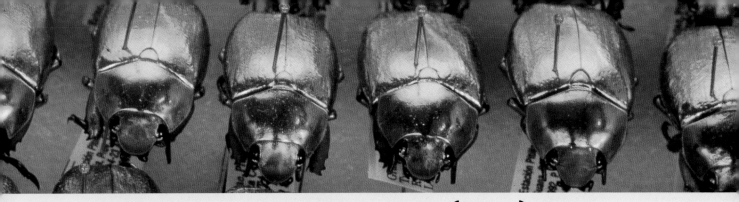

Why Biodiversity Doesn't (Yet) Pay

Foz do Iguaçu, Brazil—At the International Congress of Entomologists last summer, Ebbe Nielsen, director of the Australian National Insect Collection in Canberra, reflected on the reasons why, despite the 1992 Convention on Biological Diversity signed here in Brazil by 178 countries, so little has happened since to secure the world's threatened species. "You and I can say extinction rates are too high and we have to stop it, but to convince the politicians we have to have convincing reasons," he said. "In developing countries, the economic pressures are so high, people use whatever they can find today to survive until tomorrow. As long as that's the case, there will be no support for biodiversity at all."

Not, that is, unless it can be made more profitable to leave a forest standing or a wetland wet than it is to convert the land to farm, pasture or parking lot. Unfortunately, time has not been kind to the several arguments environmentalists have made to assign economic value to each one of perhaps 10 million species.

A Hedge against Disease and Famine

"Narrowly utilitarian arguments say: The incredible genetic diversity contained in the population and species diversity that we are heirs to is ultimately the raw stuff of tomorrow's biotechnological revolution," observes Robert May of Oxford. "It is the source of new drugs." Or new foods, adds E. O. Wilson of Harvard, should something happen to the 30 crops that supply 90 percent of the calories to the human diet, or to the 14 animal species that make up 90 percent of our livestock.

"Some people who say that may even believe it," May continues. "I don't. Give us 20 or 30 years and we will design new drugs from the molecule up, as we are already beginning to do."

Hopes were raised 10 years ago by reports that Merck had paid $1.14 million to InBio, a Costa Rican conservation group, for novel chemicals extracted from rain-forest species. The contract would return royalties to InBio if any of the leads became drugs. But none have, and Merck terminated the agreement in 1999. Shaman Pharmaceuticals, founded in 1989 to commercialize traditional medicinal plants, got as far as late-stage clinical trials but then went bankrupt. And given, as Wilson himself notes in *The Diversity of Life,* that more than 90 percent of the known varieties of the basic food plants are on deposit in seed banks, national parks are hardly the cheapest form of insurance against crop failures.

Ecosystem Services

"Potentially the strongest argument," May says, "is a broadly utilitarian one: ecological systems deliver services we're only just beginning to think of trying to estimate. We do not understand how much you can simplify these systems and yet still have them function. As Aldo Leopold once said, the first rule of intelligent tinkering is to keep all the pieces."

The trouble with this argument, explains Columbia University economist Geoffrey Heal, is that "it does not make sense to ask about the value of replacing a life-support system." Economics can only assign values to things for which there are markets, he says. If all oil were to vanish, for example, we could switch to alternative fuels that cost $50 a barrel. But that does not determine the price of oil.

And although recent experiments suggest that removing a large fraction of species from a small area lowers its biomass and ability to soak up carbon dioxide, scientists cannot say yet whether the principle applies to whole ecosystems. "It may be that a grievously simplified world—the world of the cult movie *Blade Runner*—can be so run that we can survive in it," May concedes.

A Duty of Stewardship

Because science knows so little of the millions of species out there, let alone what complex roles each one plays in the ecosystems it inhabits, it may never be possible for economics to come to the aid of endangered species. A moral argument may thus be the best last hope—certainly it is appeals to leaders' sense of stewardship that have accomplished the most so far. But is it hazardous for scientists to make it?

They do, of course, in various forms. To Wilson, "a species is a masterpiece of evolution, a million-year-old entity encoded by five billion genetic letters, exquisitely adapted to the niche it inhabits." For that reason, conservation biologist David Ehrenfeld proposed in *The Arrogance of Humanism,* "long-standing existence in Nature is deemed to carry with it the unimpeachable right to continued existence."

Winning public recognition of such a right will take much education and persuasion. According to a poll last year, fewer than one quarter of Americans recognized the term "biological diversity." Three quarters expressed concern about species and habitat loss, but that is down from 87 percent in 1996. And May observes that the concept of biodiversity stewardship "is a developed-world luxury. If we were in abject poverty trying to put food in the mouth of the fifth child, the argument would have less resonance."

But if scientists "proselytize on behalf of biodiversity"—as Wilson, Lovejoy, Ehrlich and many others have done—they should realize that "such work carries perils," advises David Takacs of California State University at Monterey Bay. "Advocacy threatens to undermine the perception of value neutrality and objectivity that leads laypersons to listen to scientists in the first place." And yet if those who know rare species best and love them most cannot speak openly on their behalf, who will?

think about in Europe, Asia and other long-settled regions. On the other hand, we may have more time than we fear to prevent future catastrophes in areas where humans have been part of the ecosystem for a while—and less time than we hope to avoid them in what little wilderness remains pristine.

"The question is how to deal with uncertainty, because there really is no way to make that uncertainty go away," Winemiller argues. "We think the situation is extremely serious; we just don't think the species extinction issue is the peg the conservation movement should hang its hat on. Otherwise, if it turns out to be wrong, where does that leave us?"

Long-Term Savings

IT COULD LEAVE conservationists with less of a sense of urgency and with a handful of weak political and economic arguments [see box on opposite page]. It might also force them to realize that "many of the species in trouble today are in fact already members of the doomed, living dead," as David S. Woodruff wrote in the *Proceedings of the National Academy of Sciences* this past May. "Triage" is a dirty word to many environmentalists. "Unless we say no species loss is acceptable, then we have no line in the sand to defend, and we will be pushed back and back as losses build," Brooks argued at the Hilo meet-

fective solution to the general problem of conserving nature."

There are still a few large areas where natural selection alone determines which species succeed and which fail. "Why not save functioning ecosystems that haven't been despoiled yet?" Winemiller asks. "Places like the Guyana shield region of South America contain far more species than some of the so-called hotspots." To do so would mean purchasing tracts large enough to accommodate entire ecosystems as they roll north and south in response to the shifting climate. It would also mean prohibiting all human uses of the land. It may not be impossible: utterly undeveloped wilderness is relatively cheap, and the population of potential buyers has recently exploded.

"It turns out to be a lot easier to persuade a corporate CEO or a billionaire of the importance of the issue than it is to convince the American public," Wilson says. "With a Ted Turner or a Gordon Moore or a Craig McCaw involved, you can accomplish almost as much as a government of a developed country would with a fairly generous appropriation."

"Maybe even more," agrees Richard E. Rice, chief economist for Conservation International. With money from Moore, McCaw, Turner and other donors, CI has outcompeted logging companies for forested land in Suriname and Guyana. In Bolivia, Rice reports, "we conserved an area the size of Rhode Is-

"It turns out to be a lot easier to persuade a corporate CEO or a billionaire of the importance of the issue than it is to convince the American public." —EDWARD O. WILSON, HARVARD UNIVERSITY

ing. But losses are inevitable, Wilson says, until the human population stops growing.

"I call that the bottleneck," Wilson elaborates, "because we have to pass through that scramble for remaining resources in order to get to an era, perhaps sometime in the 22nd century, of declining population. Our goal is to carry as much of the biodiversity through as possible." Biologists are divided, however, on whether the few charismatic species now recognized as endangered should determine what gets pulled through the bottleneck.

"The argument that when you protect birds and mammals, the other things come with them just doesn't stand up to close examination," May says. A smarter goal is "to try to conserve the greatest amount of evolutionary history." Far more valuable than a panda or rhino, he suggests, are relic life-forms such as the tuatara, a large iguanalike reptile that lives only on islets off the coast of New Zealand. Just two species of tuatara remain from a group that branched off from the main stem of the reptilian evolutionary tree so long ago that this couple make up a genus, an order and almost a subclass all by themselves.

But Woodruff, who is an ecologist at the University of California at San Diego, invokes an even broader principle. "Some of us advocate a shift from saving things, the products of evolution, to saving the underlying process, evolution itself," he writes. "This process will ultimately provide us with the most cost-ef-

land for half the price of a house in my neighborhood," and the Nature Conservancy was able to have a swath of rain forest as big as Yellowstone National Park set aside for a mere $1.5 million. In late July, Peru issued to an environmental group the country's first "conservation concession"—essentially a renewable lease for the right to *not* develop the land—for 130,000 hectares of forest. Peru has now opened some 60 million hectares of its public forests to such concessions, Rice says. And efforts are under way to negotiate similar deals in Guatemala and Cameroon.

"Even without massive support in public opinion or really effective government policy in the U.S., things are turning upward," Wilson says, with a look of cautious optimism on his face. Perhaps it is a bit early to despair after all. **SA**

W. Wayt Gibbs is senior writer.

MORE TO EXPLORE

Extinction Rates. Edited by John H. Lawton and Robert M. May. Oxford University Press, 1995.

The Currency and Tempo of Extinction. Helen M. Regan et al. in the *American Naturalist*, Vol. 157, No. 1, pages 1–10; January 2001.

Encyclopedia of Biodiversity. Edited by Simon Asher Levin. Academic Press, 2001.

The Skeptical Environmentalist. Bjørn Lomborg. Cambridge University Press, 2001.

On the Termination of Species
IN REVIEW

TESTING YOUR COMPREHENSION

1) How many mass extinctions are known from the fossil record?
   a) 1
   b) 5
   c) 11
   d) 46

2) The last mass extinction, which took place at the end of the cretaceous period and marked the decline of the dinosaurs, took place _____ years ago.
   a) 6,500
   b) 6.5 million
   c) 65 million
   d) 650 million

3) Why would "vertebrate chauvinism" undermine the ability of biologists to accurately estimate the rate of species extinction?
   a) Vertebrates rarely leave intact fossils.
   b) Vertebrates make up only a very small fraction of the total diversity of life.
   c) Vertebrates live longer than most other species.
   d) Vertebrates evolve more quickly than other species.

4) According to a frequently cited book by biologist E.O. Wilson, _____ different species become extinct every year.
   a) 27
   b) 270
   c) 27,000
   d) 2.7 million

5) Which of the following factors result(s) in an underestimation of the natural rate of species extinction (the "background rate") in the fossil record?
   a) Different species survive different lengths of time, with mammals among the shortest-lived.
   b) Most species do not become fossilized.
   c) Most species are short-lived, and a few long-lived species skew the average.
   d) All of the above

6) Approximately how many non-microbial species have been identified and named?
   a) 180,000
   b) 1.8 million
   c) 180 million
   d) 18 billion

7) Which kind of organism comes in more species than the total species of all the other multicellular organisms combined?
   a) insects
   b) fungi
   c) nematodes
   d) protozoa

8) What does the "species–area relation" say about the relationship between habitat loss and species extinction?
   a) As habitat area is lost, the number of species living in it decreases.
   b) As habitat area is lost, the number of species living in it increases.
   c) As habitat area is lost, the number of species living in it does not change.
   d) As habitat area is gained, the number of species living in it decreases.

9) In what biome do most endangered species live?
   a) chaparral
   b) deciduous forest
   c) desert
   d) tropical forest

10) Which of the following is NOT a common argument made by conservationists in favor of maintaining species diversity?

a) Most of our drugs are derived from natural species.

b) Whole ecosystems provide irreplaceable services to the biosphere.

c) Humans have a moral obligation to act as stewards for the diversity of life.

d) Human nutrition depends upon having a wide variety of natural species to eat.

BIOLOGY IN SOCIETY

1) Three arguments in favor of maintaining species diversity are listed on page 38. Which one do you find the most convincing, and why? Which one do you find the least convincing, and why? Can you think of other arguments for or against maintaining species diversity?

2) Imagine that you purchased a plot of land to build your dream house. But before you can start clearing the building site, an environmental group performs a study that purports to show that the sole remaining members of a species of bird occupies trees in and around your intended home. They seek a court injunction to prevent you from starting construction. How would you react to this situation?

3) One basic point of the conservationist movement is that the rate of species extinction is increasing. However, it is possible that biologists will never be able to truly calculate the number of species living on Earth or the number that become extinct. Given the potentially permanent uncertainties, do you think it is prudent to base public policies on the larger assumption that species are, in fact, disappearing at an increasing rate? Do you feel that the data col-

lected so far justifies drastic governmental action? Is there any evidence that you would like to see collected before deciding? Is it likely that such evidence can be collected in the near future?

THINKING ABOUT SCIENCE

1) Estimating the rate of species extinction is a very difficult task. Search this article for sources of error in such calculations. Which of the discussed sources do you think introduces the largest error? Can you imagine a way to reduce or eliminate that error?

2) The species—area relation states that as the size of a habitat falls, the number of species living in it decreases (by the 3rd root to the 6th root). Design a hypothetical experiment to test the species — area relationship. What hypothesis would form the basis of your experiment? What would be your experimental design? Prepare some hypothetical data of the type that you might collect. How would you present these data? What conclusions would your hypothetical data lead you to? What sources of error can you imagine in your experimental design? How could further research reduce those potential sources of error?

WRITING ABOUT SCIENCE

Put yourself in the shoes of a native South American who makes a living from agriculture. Your growing family requires you to expand your plot. To do this, you must burn some tropical forest. A United States agency requests you to refrain from burning because of the potential damage to rare species living in the forest. Write an essay that responds to the request and presents the competing issues.

Testing Your Comprehension Answers:
1b, 2c, 3b, 4c, 5d, 6b, 7a, 8a, 9d, 10d.

A YEAR AFTER DOCTORS BEGAN IMPLANTING THE ABIOCOR IN DYING PATIENTS, THE PROSPECTS OF THE DEVICE ARE UNCERTAIN

The Trials of an Artificial HEART

By Steve Ditlea

The permanent replacement of a failing human heart with an implanted mechanical device has long been one of medicine's most elusive goals. Last year this quest entered a crucial new phase as doctors at several U.S. hospitals began the initial clinical trials of a grapefruit-size plastic-and-titanium machine called the AbioCor. Developed by Abiomed, a company based in Danvers, Mass., the AbioCor is the first replacement heart to be completely enclosed within a patient's body. Earlier devices such as the Jarvik-7, which gained worldwide notoriety in the 1980s, awkwardly tethered patients to an air compressor. In contrast, the AbioCor does not require tubes or wires piercing the skin. In July 2001 Robert L. Tools, a 59-year-old former Marine whose heart had become too weak to pump effectively, became the first recipient of this artificial heart.

Over the next nine months, surgeons replaced the failing hearts of six more patients with the AbioCor. But the initial trials have had mixed results. As of press time, five of the seven patients had died: two within a day of the implantation procedure, one within two months, and two within five months. (Tools died last November.) One of the two survivors has lived for more than eight months with the device, the other for more than six months. Because all the patients were seriously ill to begin with—only people deemed likely to die within a month were eligible for implantation—Abiomed officials argue that the artificial heart is proving its worth. The company has acknowledged, however, that a flaw in the device's attachments to the body might have led to the formation of the blood clots that caused strokes in three of the patients.

ABIOCOR, an artificial heart made of plastic and titanium, has been in clinical trials for the past year.

With the clinical trials only a year old, it is obviously too early to say whether the AbioCor will be a breakthrough or a disappointment. If the U.S. Food and Drug Administration decides that the device shows promise, it may allow Abiomed to implant its artificial heart in patients who are not as severely ill as those in the initial group. Company officials hope that eventually the rate of survival after implantation will surpass the rate after heart transplants (about 75 percent of the recipients of donor hearts are still alive five years after the transplant). Fewer than 2,500 donor hearts become available every year in the U.S., whereas more than 4,000 Americans are on waiting lists for transplants; for many of those patients, AbioCor could be a lifesaver.

But the artificial heart is competing against less radical treatments, one of which has already proved quite successful. Doctors have been able to restore adequate cardiac function in thousands of patients by attaching a pump to the left ventricle, the chamber most likely to fail. These ventricular-assist devices were originally intended as a short-term therapy for people awaiting transplants, but recent studies show that the pumps can keep patients alive for two years or more [*see box on pages 48 and 49*]. Meanwhile other studies have overturned generations of medical wisdom by suggesting that the human heart can repair itself by generating new muscle tissue. Researchers are now racing to develop therapies using stem cells that could help the heart heal [*see box on page 50*].

Heart History

THE ORIGINS of the artificial heart go back half a century. In 1957 Willem J. Kolff (inventor of the dialysis machine) and Tetsuzo Akutsu of the Cleveland Clinic replaced the heart of a dog with a polyvinyl chloride device driven by an air pump. The animal survived for 90 minutes. Seven years later President Lyndon B. Johnson established an artificial-heart program at the National Institutes of Health. In 1969 Denton A. Cooley of the

Overview/*AbioCor Heart*

- The goal of implanting a permanent mechanical substitute for a failing human heart was all but abandoned after controversial attempts in the 1980s. The clinical trials of the AbioCor, a new artificial heart designed to be completely enclosed in a patient's body, began in July 2001.
- The trials have had mixed results so far. Of the seven severely ill patients who received the AbioCor, two died within a day of the implantation, one within two months, and two within five months. Although the artificial heart did not cause infections, three patients suffered strokes.
- If the survival rate of the AbioCor improves, it could eventually become an alternative for people on the long waiting lists for heart transplants. But the device may have to compete with less radical treatments such as ventricular-assist devices and therapies using stem cells.

LIKE A HUMAN HEART, the AbioCor has chambers for pumping blood on its left and right sides. Oxygenated blood from the lungs flows into and out of the left chamber, and oxygen-depleted blood from the body flows into and out of the right chamber. Between the chambers is the mechanical equivalent of the heart's walls: a hermetically sealed mechanism that generates the pumping motions.

At the center of this mechanism, an electric motor turns a miniaturized centrifugal pump at 5,000 to 9,000 rotations a minute. The pump propels a viscous hydraulic fluid; a second electric motor turns a gating valve that allows the fluid to alternately fill and empty from the two outer sections of the pumping mechanism. As fluid fills the left section, its plastic membrane bulges outward, pushing blood out of the AbioCor's left chamber. At the same time, hydraulic fluid empties from the right section and its membrane deflates, allowing blood to flow into the device's right chamber.

The AbioCor's four valves are made of plastic and configured like natural heart valves. The inflow conduits are connected to the left and right atria of the excised heart, and the outflow conduits are fitted to the arteries. The device weighs about one kilogram and consumes about 20 watts of power. The internal battery, electrical induction coil and controller module add another kilogram to the implanted system. Lithium-ion batteries worn on the patient's belt continuously recharge the internal battery using the induction coil. A bedside console can also be used as a power source and monitoring system.   —S.D.

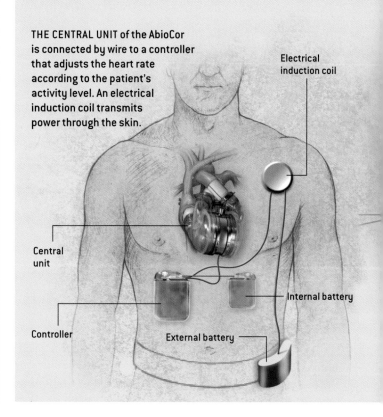

THE CENTRAL UNIT of the AbioCor is connected by wire to a controller that adjusts the heart rate according to the patient's activity level. An electrical induction coil transmits power through the skin.

Electrical induction coil

Central unit

Controller

Internal battery

External battery

HOW THE ABIOCOR WORKS

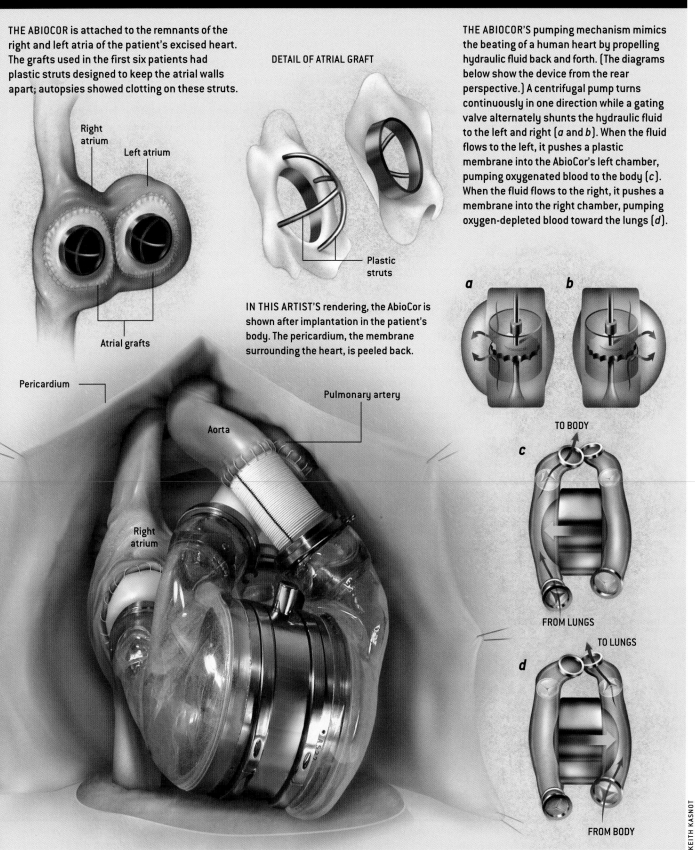

THE ABIOCOR is attached to the remnants of the right and left atria of the patient's excised heart. The grafts used in the first six patients had plastic struts designed to keep the atrial walls apart; autopsies showed clotting on these struts.

Right atrium

Left atrium

Atrial grafts

DETAIL OF ATRIAL GRAFT

Plastic struts

IN THIS ARTIST'S rendering, the AbioCor is shown after implantation in the patient's body. The pericardium, the membrane surrounding the heart, is peeled back.

Pericardium

Aorta

Pulmonary artery

Right atrium

THE ABIOCOR'S pumping mechanism mimics the beating of a human heart by propelling hydraulic fluid back and forth. (The diagrams below show the device from the rear perspective.) A centrifugal pump turns continuously in one direction while a gating valve alternately shunts the hydraulic fluid to the left and right (*a* and *b*). When the fluid flows to the left, it pushes a plastic membrane into the AbioCor's left chamber, pumping oxygenated blood to the body (*c*). When the fluid flows to the right, it pushes a membrane into the right chamber, pumping oxygen-depleted blood toward the lungs (*d*).

a *b*

c

TO BODY

FROM LUNGS

d

TO LUNGS

FROM BODY

KEITH KASNOT

The AbioCor trials revive some troubling questions

DURING THE CLINICAL TRIALS of the Jarvik-7 artificial heart, medical ethicists voiced concern about the suffering of the patients and the intense media coverage that descended on them. Now those issues have surfaced anew with the human testing of the AbioCor. So far ethicists give mixed grades to Abiomed (the maker of the device), the doctors and the press.

"The core ethical issues for the patient remain the same," says Arthur Caplan, director of the Center for Bioethics at the University of Pennsylvania School of Medicine. "First, can you get truly informed consent from a desperate, dying person? Dying is extremely coercive. There's very little you can't get a dying person to consent to." In Abiomed's favor, he rates the firm's 13-page consent form as "very strong" in terms of disclosing risks, and he commends the company's funding of independent patient advocates to inform patients and their families. But

Caplan wonders whether the right patients are enrolled in the trials: "I've argued that for some treatments it doesn't make sense to test first in the most severely ill, because you have an impossible time sorting out what's caused by the illness and what's caused by the device."

George J. Annas, a professor at the Boston University School of Public Health, contends that the consent procedure for the AbioCor "should be much more detailed about how you're going to die. No one's going to live for a long time on one of these. You have to plan for death. How is it going to happen? Who's going to make the decision and under what circumstances?" In two cases during the clinical trials, family members agreed to shut off the AbioCor's power, overriding its alarms, so a terminally failing patient could die.

Another source of controversy has been Abiomed's policy of limiting the release of information from the trials. For example, company officials will not announce a patient's identity until 30 days after an implantation (leaks at the hospital, however, have sometimes forced

them to do so sooner). Although the policy has prevented a repeat of the media frenzy surrounding the Jarvik-7 trials, some ethicists have emphasized the need for full disclosure of the medical problems encountered during the human testing. Renee Fox, a social sciences professor at the University of Pennsylvania, notes that Abiomed's reporting of negative developments has been timely, for the most part. But, she adds, "there has been a tendency by the company and the physicians to interpret adverse events as not due to the implanted heart. In each case there has been an attempt to say that this is due to the underlying disease state of the patient rather than any harm that the device may have done."

Ethicists point out that journalists have erred, too, by writing overoptimistic stories about the AbioCor. It was a hopeful cover story in *Newsweek* that convinced Robert L. Tools to volunteer for the first implant. Says Ronald Munson, a professor of philosophy of science and medicine at the University of Missouri at St. Louis, "The press shouldn't evangelize a medical procedure." —*S.D.*

Texas Heart Institute in Houston implanted an artificial heart into a person for the first time, but only as an emergency measure. The device was intended as a bridge to transplant—it kept the patient alive for 64 hours until a human heart could be found for him. (The patient received the transplant but died two and a half days later.) The next artificial-heart implant was not attempted until 1981. The patient lived for 55 hours with the bridge-to-transplant device before receiving a human heart.

Then came the most publicized clinical trials in modern medicine: cardiac surgeon William DeVries's four permanent implants of the Jarvik-7 artificial heart. When DeVries performed the first cardiac replacement in 1982 at the University of Utah Medical Center, patient Barney B. Clark became an instant celebrity. His medical status was reported almost daily. Reporters tried to sneak into the intensive care unit in laundry baskets or disguised as physicians. By the time Clark died 112 days later—from multiple organ failure after suffering numerous infections—the media had provided a detailed chronicle of the medical problems and discomfort he had experienced.

Nearly two years later DeVries performed his next Jarvik-7 implant, this time at Norton Audubon Hospital in Louisville, Ky., on patient William Schroeder. Schroeder survived on the artificial heart for 620 days, the longest of anyone to date, but

it took a tremendous toll on him: strokes, infections, fever and a year of being fed through a tube. The third Jarvik-7 recipient lived for 488 days, and the fourth died after just 10 days. Although several hospitals successfully used a slightly smaller version of the Jarvik-7 as a bridge-to-transplant device for hundreds of patients, most medical professionals abandoned the idea of a permanent artificial heart.

But an engineer named David Lederman believed that the concept still held promise. Lederman had worked on developing an artificial heart at the medical research subsidiary of Avco, an aerospace company, and in 1981 he founded Abiomed. He and his colleagues closely followed the clinical trials of the Jarvik-7 and considered ways to improve it. The external air compressor that powered the device was bulky and noisy. Infectious bacteria could easily lodge where the tubing pierced the patient's skin. And inside the heart itself were surface discontinuities where platelets and white blood cells could coagulate into a thrombus, a solid clot that could circulate in the blood and lodge in the brain, causing a stroke.

In 1988 the National Heart, Lung and Blood Institute at the NIH decided to cut off support for replacement-heart research and instead channel funds to ventricular-assist pumps. Lederman went to Washington along with representatives from other research teams to lobby against the change. They convinced a group of senators from their home states to help restore NIH support, resuscitating research programs at two universities (Utah and Pennsylvania State) and two companies (Nimbus in Rancho Cordova, Calif., and Abiomed). Today Abiomed is the last artificial-heart developer left from that group. The company has received nearly $20 million in federal research grants. Its government funding ended in 2000, but that same year Abiomed raised $96 million in a stock offering.

Lederman and his colleagues are doggedly pursuing a medical technology whose time others believe may have already come and gone. In the conference room at Abiomed's headquarters in an office park north of Boston, Lederman attributes his firm's tenacity to its team of researchers: "No one else had the commitment to say there is no alternative to success. This is important stuff. I take pride in the fact that we took it so seriously." It is also evident that for Lederman this is a personal matter: in 1980 his father died suddenly of a heart attack.

Designing AbioCor

THE ABIOCOR is not powered by an air compressor as the Jarvik-7 was. Hidden behind the device's central band of metal is the heart of this heart: a pair of electric motors driving a pump-and-valve system. This pumping mechanism propels hydraulic fluid back and forth, causing a pair of plastic membranes to beat like the inner walls of a human heart [see box on pages 44 and 45].

But this innovation was only the start. To be truly self-contained, the device needed a small, implantable controller that could vary the heart rate to match the patient's activity level. The controller developed by Abiomed is the size of a small paperback; implanted in the patient's abdomen, it is connected to the artificial heart by wire. Sensors inside the heart measure the pressure of the blood filling the right chamber—the blood returning to the heart from the body—and the controller adjusts the heart rate accordingly. The rate can range from 80 to 150 beats a minute. If the clinical trials show that this control system is adequate, it could be shrunk down to a single microchip that would fit on the AbioCor's central unit.

Abiomed also developed a way to power the artificial heart's motors without the use of skin-penetrating wires, which can leave the patient prone to infections. An internal battery implanted in the patient's abdomen can hold enough charge to sustain the heart for 20 minutes. This battery is continuously recharged through electromagnetic induction—the same process used in electric toothbrushes. The internal battery is wired to a passive electrical transfer coil under the patient's skin. Another coil outside the patient's skin, wired to an external battery, transmits power through the skin tissue with minimal radiation and heat. The patient can wear the external battery on a belt, along with a separate monitor that alerts the patient if the battery's charge runs low.

A major concern was to design the AbioCor so that it could pump blood without creating clots. When Lederman had worked for Avco, he had conducted four years of research on the interaction between blood and synthetic materials, studying

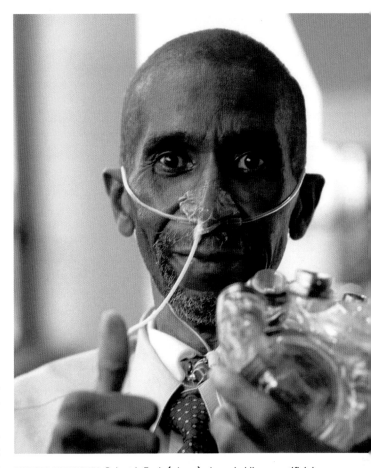

ABIOCOR RECIPIENTS: Robert L. Tools (*above*), shown holding an artificial heart like the one in his chest, became the first AbioCor patient in July 2001. He died in November after suffering a severe stroke. The second recipient, Tom Christerson (*below*), has lived the longest with the AbioCor—more than eight months as of press time. Shown here with a physical therapist at Jewish Hospital in Louisville, Ky., Christerson returned home in April.

the reaction rates of various coagulation processes. Essentially the AbioCor minimizes clotting by making sure that the blood cells do not have time to stick together. Blood flows swiftly through the device, and there are no areas where pooling can occur. All the surfaces of the device that are in contact with blood are made of Angioflex, a biologically inert polyurethane plastic. The contact surfaces are also extremely smooth because clots can form on irregular surfaces. Says Lederman, "We had to make a system that was totally seamless."

Trial and Error

AFTER TESTING its artificial heart in calves and pigs, Abiomed received permission from the FDA in January 2001 to begin clinical trials in humans. The FDA would determine the success of the trials by reviewing the patients' survival rates and quality of life, as measured by standard assessment tests. Only patients who were ineligible for a heart transplant could volunteer for the implantation. The size of the AbioCor also ruled out certain patients: the device can fit inside the chests of only half of adult men and 18 percent of adult women. (Abiomed is developing a smaller, second-generation heart that would fit most men and women.) For each procedure, Abiomed agreed to pay for the device and its support. Hospitals and doctors participating in the trials would donate facilities and care. The total cost of each implantation and subsequent treatment: more than $1 million.

On July 2, 2001, the first AbioCor was implanted in Robert L. Tools at Jewish Hospital in Louisville, Ky., by surgeons Laman A. Gray, Jr., and Robert D. Dowling in a seven-hour operation. Tools had been suffering from diabetes and kidney failure as well as congestive heart failure. Before the heart replacement, he could barely raise his head. After the procedure, Tools experienced internal bleeding and lung problems, but within two months his kidney function had returned to normal and he had enough strength to be taken on occasional outings from the hospital. His doctors hoped he would be able to go home by Christmas. Tools's bleeding problems persisted, however, making it difficult for doctors to administer the anticoagulant drugs intended to prevent clot formation. On November 11 he suffered a severe stroke that paralyzed the right side of his body. He died 19 days later from complications following gastrointestinal bleeding.

The second recipient of the AbioCor, a 71-year-old retired businessman named Tom Christerson, has fared much better so far. Surgeons at Jewish Hospital implanted the device in Christerson on September 13, 2001. After a steady recovery, he left the hospital in March to take up residence in a nearby hotel, where he and his family could learn how to tend to the artificial heart on their own. The next month he returned to his home in Central City, Ky. In the following weeks, Christerson continued his physical therapy and visited Jewish Hospital for weekly checkups. His car was wired so that he could use it as a power source for his artificial heart.

At the Texas Heart Institute, O. H. "Bud" Frazier—the surgeon who has the record for performing the most heart transplants—implanted the AbioCor into two patients. One lived with the device for more than four months before dying of com-

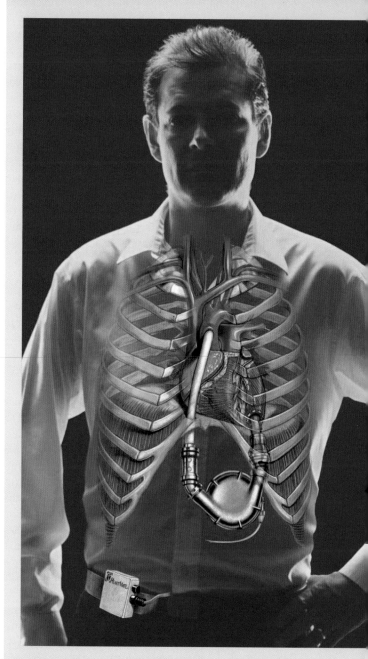

HEARTMATE PUMP, the most widely used ventricular-assist device, is implanted in a patient's abdomen, as shown in this artist's rendering. Attached to a failing left ventricle, the device pumps oxygenated blood to the body.

plications from a stroke; the other died within a day of the implantation, succumbing to uncontrolled bleeding after spending 20 hours on the operating table. Implantations have also been performed at the University of California at Los Angeles Medical Center and Hahnemann University Hospital in Philadelphia. The Los Angeles patient lived for a little less than two months before heart support was withdrawn following multiple organ

Ventricular-assist devices emerge as an alternative to heart replacement

IN NOVEMBER 2001, soon after human testing of the AbioCor began, researchers reported that another clinical trial had demonstrated the benefits of a less drastic treatment for heart failure. The left ventricular assist device (LVAD)—a pump implanted in the chest or abdomen and attached to the heart's left ventricle, the chamber that pumps oxygenated blood to the body—had been developed as a short-term therapy for patients awaiting heart transplants. But the trial showed that LVADs can keep patients alive for two years or more, and the Food and Drug Administration is expected to approve the devices for long-term use.

The study evaluated 68 patients with implants of the HeartMate, the most widely used LVAD, and 61 patients who received medical therapy, including potent cardiac drugs. After a year, more than half of those with LVADs were still alive, compared with only one quarter of those on medical therapy. At two years, the survival rates were 23 percent for the LVAD group and 8 percent for the medical group. The longest stint on the HeartMate is now more than three years; the longest survivor of the medical group died after 798 days. "There are still 21 patients ongoing with the devices," notes Eric Rose, surgeon in chief at Columbia Presbyterian Medical Center in New York City and principal investigator for the trial. "This sets a new benchmark for treating the disease."

The HeartMate, made by Thoratec in Pleasanton, Calif., is far from perfect. Many of the implanted test subjects suffered serious infections because the device is connected to an external battery by a skin-piercing tube. Other HeartMate patients died from mechanical malfunctions such as motor

JARVIK 2000 is the only assist device small enough to fit within the left ventricle. Robert K. Jarvik, inventor of the Jarvik-7 artificial heart, now develops assist devices rather than replacement hearts.

failure. But Thoratec has already improved on the current version of the device and is developing second- and third-generation systems designed to last eight and 15 years, respectively.

Another LVAD, called the LionHeart, made by Arrow International in Reading, Pa., is a fully implantable system with no skin-piercing tubes or wires. Now in clinical trials, the LionHeart uses an electrical induction coil like the AbioCor's to transmit power through the skin. The MicroMed DeBakey VAD is also fully implantable, but it propels blood in a steady flow rather than pumping it like a natural heart. Proponents of this technology tout its efficiency and reliability; critics argue that a pulsating heartbeat is needed to keep blood vessels clear. Cardiac pioneer Michael E. DeBakey, who performed the first successful coronary bypass in 1964, developed the device in collaboration with one of his patients, David Saucier, a NASA engineer

who had had heart transplant surgery.

Robert K. Jarvik, inventor of the Jarvik-7 artificial heart and now CEO of New York City–based Jarvik Heart, has introduced the Jarvik 2000, the only assist device small enough to be lodged inside the left ventricle. Like the DeBakey VAD, the Jarvik 2000 pumps blood in a steady flow. The device is currently in trials for bridge-to-transplant use and has been implanted in some patients for long-term use as well. Jarvik believes the device could help a less severely damaged heart to repair itself, perhaps in combination with stem cell treatments [see box on next page]. Another potential combination therapy might be the use of LVADs with the steroid clenbuterol to strengthen the heart. In a test reported last year, Magdi Yacoub of Harefield Hospital in London administered clenbuterol to 17 patients with implanted LVADs. In five of the patients, the hearts recovered enough to allow the removal of the LVADs.   —S.D.

COURTESY OF TEXAS HEART INSTITUTE

failure. The Philadelphia patient, 51-year-old James Quinn, received the AbioCor on November 5, 2001. Although he suffered a mild stroke in December, the next month he was discharged from the hospital to a nearby hotel. This past February, however, he was readmitted to the hospital with breathing difficulties. Doctors treated him for pneumonia, which became life-threatening because his lungs were already weakened by chron-

ic emphysema and pulmonary hypertension. Quinn was placed on a ventilator to help him breathe, but his recovery was slow. By mid-May, though, his condition was improving, and doctors began to wean him from the ventilator.

In January, Abiomed reported preliminary findings from the clinical trials at a press conference. Lederman noted that the artificial heart had continued to function under conditions that

Stem cells may prove to be the best medicine for injured hearts

EVERY SO OFTEN, unexpected findings turn scientific wisdom upside down. Two studies recently published in the *New England Journal of Medicine* have refuted the long-held notion that the human heart cannot repair itself after a heart attack or other injury. The research indicates that new muscle cells can indeed grow in adult hearts and that they may arise from stem cells, the undifferentiated building blocks of the body. The discovery may pave the way for therapies that encourage natural healing.

Research teams at the New York Medical College (NYMC) in Valhalla, N.Y., and the University of Udine in Italy conducted the iconoclastic experiments. The first study found chemical markers indicating new growth of muscle cells in heart samples taken from patients who had died four to 12 days after a myocardial infarction (the medical term for a heart attack). The second study, which involved the postmortem examination of female hearts transplanted into men, showed the presence of stem cells with Y chromosomes in the donated hearts. Although these stem cells could have migrated from the male recipient's bone marrow, they could have also come from the cardiac remnant to which the female heart was attached.

"Our paper suggests the possibility that cardiac stem cells may exist," says Piero Anversa, director of the Cardiovascular Research Institute at the NYMC. "We need to determine all the characteristics that prove that we are

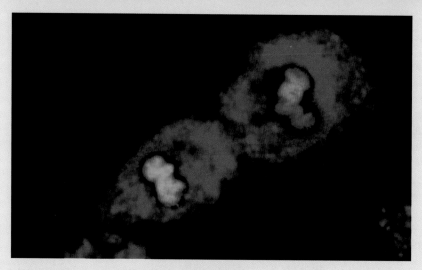

HEART MUSCLE CELL, or myocyte, is shown dividing in this microscope image of tissue taken from a patient who died shortly after a heart attack. The evidence suggests that, contrary to prevailing medical opinion, new myocytes can grow to replace damaged ones.

dealing with a primitive cell in the heart. And then we need to see whether we can mobilize these cells in areas of heart damage to promote repair."

Other medical researchers are pursuing regenerative cardiac therapies with stem cells taken from other parts of the body. Philippe Menasché, professor of cardiovascular surgery at the Bichat-Claude Bernard Hospital in Paris, has injected primitive muscle cells from patients' legs into damaged areas of their hearts during cardiac bypass surgery. Initial results from the clinical trials have been encouraging, showing thickening of heart muscle walls with functional tissue. But Menasché is cautious about therapeutic outcomes. "At best, these cells may help enhance other treatments," he says. "Imagining that you'll be able to completely regenerate an

infarcted heart is probably unrealistic."

But some biotechnology firms are entertaining even wilder hopes. Advanced Cell Technology, the Worcester, Mass.–based company that gained notoriety last year with its human cloning experiments, has already turned stem cells into beating heart cells and is trying to create transplantable patches for repairing larger areas of damage. "Eventually we want to engineer a full heart," says Robert Lanza, the company's vice president for medical and scientific development. The task would require generating cardiac muscle and blood vessel tissue as well as fabricating a dissolvable biological scaffolding material for building the heart. How far off is a biological artificial heart? According to Lanza, "We could produce a functioning heart in 10 years, with clinical trials in maybe 15 years." —*S.D.*

"EVIDENCE THAT HUMAN CARDIAC MYOCYTES DIVIDE AFTER MYOCARDIAL INFARCTION," BY A. P. BELTRAMI ET AL. IN THE *NEW ENGLAND JOURNAL OF MEDICINE*, VOL. 344, NO. 23; JUNE 7, 2001. COPYRIGHT © 2001 MASSACHUSETTS MEDICAL SOCIETY. ALL RIGHTS RESERVED

could have damaged or destroyed a natural heart, such as a severe lack of oxygen in the blood and a fever of 107 degrees Fahrenheit. Also, no patient had suffered an infection related to the device. But Abiomed acknowledged a design flaw in the artificial heart's connections to the body. The AbioCor is attached to remnants of the atria of the patient's excised heart; autopsies on two patients had shown clotting on the plastic struts of thimble-size "cages" that were intended to maintain the separation of the remaining atrial walls [*see illustration on page 45*]. Because these clots could cause strokes, Abiomed declared that

it would no longer use the plastic cages when implanting the AbioCor. The cages were needed to test the device in calves but are unnecessary in humans.

In early April, Abiomed announced that it would not be able to meet its original schedule of implanting the AbioCor in 15 volunteers by the end of June. The company said that it wanted to devote further study to its first six cases. But a week later doctors at Louisville's Jewish Hospital performed another implantation, the first using an AbioCor without the plastic cages. The artificial heart functioned properly, but the 61-year-old patient died within hours of the procedure after a clot lodged in his lungs. According to Laman Gray, who performed the operation with colleague Robert Dowling, the clot did not originate in the AbioCor.

The surgeons who have worked with the AbioCor remain convinced of the device's potential, despite the recent setbacks. Frazier of the Texas Heart Institute believes the formation of clots in the AbioCor's plastic cages was a complication that could not have been anticipated. "Fortunately, this one can be corrected," he says. "It's not something inherently wrong in the device." Gray concurs: "In my opinion, it's very well designed and is not thrombogenic at all. The problem has been on the inflow cage. I'm truly amazed at how well it has done in initial clinical trials." (Both surgeons consulted on the AbioCor's design and were responsible for much of its testing in animals.)

But not everyone is as sanguine as Frazier and Gray. "Total heart replacement by mechanical devices raises a number of questions that have not been addressed in this small group of patients," says Claude Lenfant, director of the National Heart, Lung and Blood Institute. "What quality of life can a total-heart-replacement patient expect? Will there be meaningful clinical benefits to the patient? Is the cost of this therapy acceptable to society?" And Robert K. Jarvik, the developer of the Jarvik-7 device that made headlines 20 years ago, now argues that permanent artificial hearts are too risky. "Cutting out the heart is practically never a good idea," he says. "It was not known in 1982 that a heart can improve a lot if you support it in certain very common disease states. That's why you should cut out the heart only in the most extreme situations."

Heart of the Matter

AS THE ABIOCOR TRIALS continue, the most crucial objective will be reducing the incidence of strokes. Doctors had originally hoped to guard against this risk by prescribing low levels of anticoagulant drugs, but some of the test subjects were so severely ill that they could not tolerate even these dosages. Because these patients had medical conditions that made them susceptible to internal bleeding, determining the best dosage of anticoagulants became a delicate balancing act: giving too much might cause the patient to bleed to death, and giving too little might cause a stroke.

Despite the need for more refinement, Lederman is satisfied with the clinical results to date. The initial goal of the trials was to show that AbioCor could keep the patients alive for at least 60 days, and four of them surpassed that mark. Says Lederman, "If most of the next patients go the way the first ones have gone but without unacceptable complications such as strokes, we plan

MORE TO EXPLORE

More information about Abiomed, the manufacturer of the AbioCor, is available at **www.abiomed.com**

The Web site of the Implantable Artificial Heart Project at Jewish Hospital in Louisville, Ky., is **www.heartpioneers.com**

The Texas Heart Institute in Houston: **www.tmc.edu/thi**

The National Heart, Lung and Blood Institute at the National Institutes of Health: **www.nhlbi.nih.gov/index.htm**

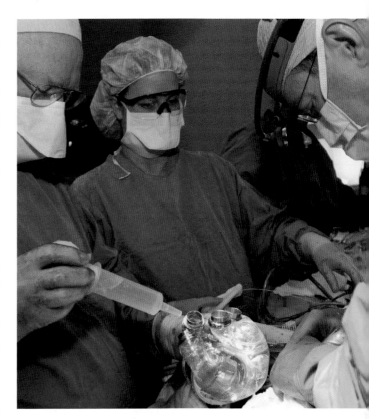

JUST BEFORE IMPLANTING the AbioCor into patient Tom Christerson, surgeons Laman A. Gray, Jr. (*left*), and Robert D. Dowling (*right*) evacuate the air from the artificial heart to prevent blood clotting. The procedure was performed on September 13, 2001, at Louisville's Jewish Hospital.

to ask the FDA to authorize clinical use of the system for patients who are on their last breath. We think we have a convincing argument that we can give patients with less than 30 days to live many months of quality life." But some medical ethicists have questioned this approach, saying that people at death's door might consent to any procedure, no matter what the consequences [*see box on page 46*].

And then there is the issue of how to define an acceptable quality of life. In 1981 Jarvik wrote that for the artificial heart to achieve its goal "it must be forgettable"—that is, the device should be so unobtrusive and reliable that patients would be able to ignore it [see "The Total Artificial Heart," by Robert K. Jarvik; SCIENTIFIC AMERICAN, January 1981]. Does the AbioCor meet that standard? Tools's wife, Carol, says that her husband was aware that his old heartbeat had been replaced by the AbioCor's low, steady whir. "Sometimes he'd lie there, and he would listen to it," she says. "But other times he would forget it.... [He] always knew it was there, because he still had to power it. It's not like replacing a hip." Still, she believes that the quality of life during his last months was good: "He had a chance to live quite well, although unfortunately, it was shorter than we would have liked." She adds, "He never had any regrets about it." SA

Steve Ditlea is a freelance journalist based in Spuyten Duyvil, N.Y. He has been covering technology since 1978.

The Trials of an Artificial Heart
IN REVIEW

TESTING YOUR COMPREHENSION

1) How does the AbioCor artificial heart, introduced in 2001, differ from older models of artificial hearts?
   a) It can completely replace heart functions.
   b) It can be completely enclosed within a patient's body.
   c) It is the first to significantly extend life spans.
   d) It is made from stem cells.

2) Which of the following statements best describes the results achieved so far in implanting AbioCor artificial hearts into living human patients?
   a) None of the patients benefited from the procedure.
   b) Some of the patients benefited by living months longer than they would have otherwise.
   c) Some of the patients benefited by living years longer than they would have otherwise.
   d) All of the patients benefited by living longer than they would have otherwise.

3) What is the survival rate among heart transplant recipients?
   a) 25%
   b) 50%
   c) 75%
   d) 100%

4) The human heart uses muscle contractions to propel blood. What mechanism does the AbioCor artificial heart use?
   a) artificially stimulated muscle contractions
   b) magnetic impulses transmitted through the skin
   c) air compressors
   d) the movement of hydraulic fluid

5) Why is it important that an artificial heart be coated with smooth, non-biologically reactive surfaces?
   a) to ease the flow of blood through the heart
   b) to decrease the amount of power required to run the pump
   c) to reduce the risk of immune system rejection
   d) to prevent the formation of clots

6) Which chamber of the heart pumps blood to the body?
   a) left atrium
   b) left ventricle
   c) right atrium
   d) right ventricle

7) A less drastic treatment than an artificial heart, a(n) _____ is a pump that assists the heart with its natural pumping action.
   a) arterial control device
   b) atrial control devise
   c) ventricular-assist device
   d) ventricular-replacement device

8) What is the advantage of using an external battery with an induction coil to power an artificial pump over an external battery connected by a wire?
   a) The induction coil has no wires entering the body, reducing risk of infection.
   b) The induction coil can stay charged for longer.
   c) The induction coil uses no artificial parts.
   d) The induction coil is less likely to cause strokes.

9) What is the greatest medical problem with the current generation of artificial hearts?
   a) insufficient blood flow
   b) clot formation
   c) large size
   d) mechanical failure

10) Which of the following recent discoveries may contribute to future cardiac therapies?
   a) genetic engineering of heart cells to replace defective genes
   b) embryonic stem cells
   c) evidence of new growth of muscle cells in adult hearts and cardiac stem cells
   d) transplantable patches of beating heart cells created from stem cells

BIOLOGY IN SOCIETY

1) As discussed in this article, ethicists are troubled by the idea of receiving "informed consent" from patients with just a few months to live. What is "informed consent?" How is it typically obtained? Why is obtaining informed consent more difficult from critically ill patients? How strongly do you feel about these moral questions? Would your opinion be different if you or a loved one were the patient?

2) Given the results presented in this article, would you be willing to receive an AbioCor artificial heart? Why or why not?

3) If the safety of the AbioCor is improved, do you think it is ethical to spend up to $1 million to implant an artificial heart into a single patient? Should insurance companies be obligated to pay? If not, should wealthy patients be the only potential recipients of an artificial heart?

THINKING ABOUT SCIENCE

1) This article compares the performance of artificial hearts versus ventricular-assist devices. Design a hypothetical study intended to determine which of these devices is better for patients. How confident are you in the reliability of the results that would be obtained from your study? What sources of error are there within your study? Can you correct for them?

2) Before a new medical device is used in humans, it must be rigorously tested to ensure its safety and reliability. How do you think this can be done on a device like the AbioCor that can only be truly tested in humans? What special problems are associated with human clinical trials? What special precautions must be taken?

WRITING ABOUT SCIENCE

Write an essay that discusses the benefits and drawbacks of receiving an AbioCor artificial heart from the perspectives of the following people: the patient, the patient's family, the patient's doctor, the surgeons performing the implantation.

EDITOR'S NOTE:

This article is reproduced exactly as it was originally published in July 2002. Since then, there have been several significant developments regarding the AbioCor artificial heart. Most significantly, all of the seven patients who received the AbioCor artificial heart are now deceased.

- Patient Tom Christerson (shown in the bottom half of the figure on page 47) died on February 7, 2003. Mr. Christerson survived 17 months after his transplant, longer than anyone else with an AbioCor artificial heart. He died after an internal membrane in the artificial heart wore out.

- Patient James Quinn (described on page 49) died from a stroke on August 26, 2002, after surviving almost ten months with the artificial heart. After his death, Mr. Quinn's widow sued the heart's manufacturer and others, claiming that her husband's quality of life was not what had been promised. In June 2003, the parties settled the lawsuit for $125,000.

For current information on clinical trials and other news regarding the AbioCor artificial heart, you can visit the manufacturer's website at http://www.abiomed.com/home.html.

Testing Your Comprehension Answers:
1b, 2b, 3c, 4d, 5d, 6b, 7c, 8a, 9b, 10c.

FAN-SHAPED LEAVES of the ginkgo tree are shown in this drawing from *Flora Japonica*, a book written by 19th-century German physician Philipp Franz von Siebold. The extract obtained from the leaves (*inset on opposite page*) is one of the most widely used herbal treatments aimed at improving memory.

THE LOWDOWN ON GINKGO BILOBA

This popular herbal supplement may slightly improve your memory, but you can get the same effect by eating a candy bar

By Paul E. Gold, Larry Cahill and Gary L. Wenk

FROM *FLORA JAPONICA*, BY SIEBOLD AND ZUCCARINI, LEIDEN 1835/42, IN THE LEIDEN UNIVERSITY BRANCH OF THE NATIONAL HERBARIUM OF THE NETHERLANDS (*opposite page*); GETTY IMAGES (*above*)

T he ginkgo tree (*Ginkgo biloba*) is remarkable in many ways. Although indigenous to Korea, China and Japan, the tree can be found in parks and along city sidewalks around the world. It may grow as high as 40 meters and live for more than 1,000 years. Ginkgo fossils have been dated as far back as 250 million years ago, and Charles Darwin referred to the tree as "a living fossil." Nowadays, however, the ginkgo's primary claim to fame is the extract obtained from its fan-shaped leaves.

The use of ginkgo leaf extracts can be traced back for centuries in traditional Chinese medicine. Today ginkgo biloba is perhaps the most widely used herbal treatment aimed at augmenting cognitive functions—that is, improving memory, learning, alertness, mood and so on. Ginkgo is especially popular in Europe; officials in Germany recently approved the extract for treating dementia. In the U.S. the National Institute on Aging is currently supporting a clinical trial to evaluate the efficacy of ginkgo in treating the symptoms of Alzheimer's disease.

But is there any evidence that ginkgo biloba can really improve cognitive functions? Information on most dietary supplements is based far more on folklore than on experimental findings. Because the U.S. Food and Drug Administration does not regulate herbal treatments, the manufacturers are not required to test the effectiveness or safety of their products. More attention to supplements such as ginkgo biloba is clearly warranted; even if the products do not cause medical problems, they can be costly and may prevent patients from seeking more pragmatic treatments. In an attempt to close the gap in our knowledge, we have reviewed the experimental evidence both for and against the usefulness of ginkgo biloba in enhancing brain functions.

Many Studies, Few Answers

THE TYPICAL DAILY DOSE of ginkgo biloba—and the one used in many of the experiments described in this article—is 120 milligrams of dried extract in two or three oral doses. The extract contains several flavonoids, a large group of natural plant products that are characterized by a specific chemical structure containing a series of carbon rings. Ginkgo extract also contains some biflavonoids, a related group of compounds, and two different types of terpenes, a class of naturally occurring chemicals that includes the active ingredients in catnip and marijuana.

Ginkgo's Effects on the Brain

RESEARCHERS CANNOT SAY FOR CERTAIN WHETHER GINKGO BILOBA CAN IMPROVE COGNITIVE FUNCTIONS, BUT THEY HAVE FOUND THAT THE EXTRACT CAN AFFECT THE BRAIN IN SEVERAL WAYS

CIRCULATORY

- Stimulates widening of the blood vessels, which leads to increased blood flow to the brain and lowered blood pressure (perhaps reducing the risk of stroke).
- Reduces cholesterol levels in the blood (excessive cholesterol is correlated with an increased risk of Alzheimer's disease).
- Inhibits the aggregation of blood platelets and the formation of clots. This may lower the risk of an occlusive stroke (caused by a clot blocking a blood vessel in the brain) but raise the chance of a hemorrhagic stroke (caused by bleeding in the brain).

ANTIOXIDANT

- Curbs the creation of free radicals, highly reactive oxygen molecules that may injure neurons and cause age-related changes in the brain.
- Alleviates the effects of cerebral ischemia—the loss of blood flow to the brain—by inhibiting the production of toxic free radicals after an ischemic episode.

GLUCOSE UTILIZATION

- Boosts the absorption of glucose, the body's primary fuel, in the frontal and parietal cortex, areas of the brain important for processing sensory information and for planning complex actions.
- Also increases glucose absorption in the nucleus accumbens and the cerebellum, brain regions involved in experiencing pleasure and controlling movement, respectively.

NEUROTRANSMITTER SYSTEMS

- Appears to help neurons in the forebrain absorb the nutrient choline

BRAIN REGIONS AFFECTED — Cerebral arteries, Frontal cortex, Parietal cortex, Nucleus accumbens, Occipital cortex, Hippocampus, Cerebellum

from the blood. Choline is one of the components of acetylcholine, a brain chemical that transmits signals between certain neurons.
- Slows the attrition of neuron receptors that direct the response to serotonin, a neurotransmitter that reduces stress and anxiety.
- Enhances the release of gamma-amino butyric acid (GABA), another neurotransmitter that can relieve anxiety. Lowering stress may reduce the level of glucocorticoid hormones in the blood, which in turn may protect the hippocampus, a brain structure critical to normal learning.
- Raises the production of norepinephrine, yet another neurotransmitter. Enhanced activation of the norepinephrine system by certain antidepressants has been shown to reduce the symptoms of depression.

To date, dozens of investigations have examined the cognitive effects of ginkgo in humans, but many of the research reports are in non-English publications or in journals with very restricted distribution, making assessment of the findings difficult. The great majority of studies have involved subjects with mild to moderate mental impairment, usually a diagnosis of early Alzheimer's. Most of the experiments that show evidence of cognitive enhancement in Alzheimer's patients have used a standardized ginkgo extract known as EGb 761.

The ginkgo researchers have usually employed tests of learning and memory; less attention has been paid to other mental functions such as attention, motivation and anxiety. Moreover, because most of the investigators introduced the tests to the subjects after long-term use of ginkgo biloba (typically several months), it is hard to identify which cognitive abilities have been affected. For example, higher scores on the memory and learning tests might stem from the fact that subjects who used ginkgo paid better attention to the initial instructions. To get more specific data on ginkgo's effects, researchers need to administer the tests both before and after the subjects take the supplement.

Because the studies have varied so greatly in the numbers of subjects and the control over experimental conditions, it is useful to focus on only the most rigorous investigations. In 1998 Barry S. Oken of Oregon Health Sciences University and his colleagues considered more than 50 studies involving subjects with mental impairment and selected four that met a conservative set of criteria, including sufficient characterization of the Alzheimer's diagnosis, use of a standardized ginkgo extract, and a placebo-controlled, double-blind design (in which neither the subjects nor the investigators know until the end whether a given patient is receiving the extract or the placebo). Each of these studies showed that the Alzheimer's patients who received ginkgo performed better on various cognitive tests than did patients who received a placebo. Improvements were evident in standardized tests measuring attention, short-term memory and reaction time; the average extent of improvement resulting from ginkgo treatment was 10 to 20 percent.

Oken and his colleagues reported that ginkgo's effect was comparable to that of the drug donepezil, which is currently the treatment of choice for Alzheimer's. Donepezil enhances brain activity by inhibiting the breakdown of acetylcholine, a brain chemical that transmits signals between certain neurons. Despite these apparently encouraging findings, though, another recent, large and well-controlled trial of EGb 761 (sponsored by its manufacturer, Dr. Willmar Schwabe Pharmaceuticals in Karlsruhe, Germany) involving patients with a mild or moderate stage of dementia reported no "systematic and clinically meaningful effect of ginkgo" on any of the cognitive tests employed.

A critical question concerns whether the ginkgo treatment in studies showing positive effects actually improved cognitive abilities in Alzheimer's patients or merely slowed their deterioration. Two different answers to this key question have come from a 1997 investigation led by Pierre L. Le Bars of the New York Institute for Medical Research. In this study, which was

University of Leeds in England administered a battery of tests to eight healthy subjects (ages 25 to 40) after they took the ginkgo extract EGb 761. Hindmarch reported that the highest dose tested (600 milligrams) improved performance in only a short-term memory test. More recently, two reports from Cognitive Drug Research, a laboratory in Reading, England, provide minor support for the view that ginkgo may enhance cognitive functions in young people. One study reported that subjects who took a dose of ginkgo performed better on tasks assessing attention than did subjects who took a placebo. The other study showed an improvement in memory among middle-aged subjects (ages 38 to 66) who were treated with a combination of ginkgo and ginseng, another herbal remedy touted as a memory aid. The effects seen in the latter study, however, could not

Information on most dietary supplements is based far more on folklore than on EXPERIMENTAL FINDINGS.

one of the four analyzed by Oken, the results varied according to the cognitive test that was employed. Measured by the Alzheimer's Disease Assessment Scale Cognitive Subscale, the performance of the patients treated with the placebo slowly deteriorated over a year, whereas the performance of patients treated with ginkgo remained stable. But according to a second test—the Geriatric Evaluation by Relative's Rating Instrument—ginkgo-treated subjects improved by about the same amount that placebo-treated subjects deteriorated.

Furthermore, at least one study has reported positive effects on mentally impaired subjects after just a single treatment of ginkgo. Herve Allain of the University of Haute Bretagne in Rennes, France, gave one fairly high dose of ginkgo—320 or 600 milligrams—to a small group of elderly people with mild, age-related memory impairment. An hour after the treatment, Allain tested the subjects' memory by rapidly presenting short lists of words or drawings and then asking the patients to recall the lists immediately afterward. Their ability to recall the rapidly presented material increased significantly after ingestion of ginkgo. This finding raises the possibility that short-term, rather than long-term, biological actions provide the basis for ginkgo's reported effects on cognition.

It should be noted that ginkgo has also been shown to impair performance. For example, in a small study of elderly people with mild to moderate memory impairment, Gurcharan S. Rai of Whittington Hospital in London and his team found that after 24 weeks of treatment, patients who took ginkgo could not recall digits as well as patients who received a placebo.

Help for the Healthy?

UNFORTUNATELY, FAR FEWER studies have examined the cognitive effects of ginkgo biloba on healthy young adults. In one small study during the mid-1980s, Ian Hindmarch of the

be attributed to ginkgo alone and did not increase with the dosage, which would be expected of a truly effective substance.

For most pharmaceuticals, researchers conduct a large number of studies with lab animals before they test the drugs in humans. Such experiments can be useful in determining a drug's safety and effectiveness. But because ginkgo is unregulated, its manufacturers have not been required to perform animal tests. As a result, there are relatively few research reports in refereed journals examining the efficacy of ginkgo in improving learning and memory in animals. Perhaps the most notable example is a 1991 study of young adult mice that were trained to press a lever to receive food. Mice treated with ginkgo for four to eight weeks learned the task slightly more quickly than did the control mice. As with humans, though, it is difficult to pin down whether ginkgo actually enhances the learning process or whether it has other effects that improve the animals' performance at a specific task. For instance, investigators have reported that repeated administration of ginkgo reduced stress in rats, and altered stress responses can themselves have a great influence on learning and memory.

If ginkgo can really enhance mental processes, how does it work? Studies of humans and lab animals have indicated several classes of biological effects that might contribute to ginkgo's putative improvement of cognitive functions [*see box on opposite page*]. Whatever its effects, ginkgo appears to pose few health

THE AUTHORS

PAUL E. GOLD, LARRY CAHILL and GARY L. WENK are leading authorities on the enhancement of brain functions. Gold is professor of psychology and neuroscience at the University of Illinois at Urbana-Champaign. Cahill is assistant professor of neurobiology and behavior and fellow of the Center for the Neurobiology of Learning and Memory at the University of California, Irvine. Wenk is professor of psychology and neurology at the University of Arizona.

MONICA STEVENSON *FoodPix*

The Other "Brain Boosters"

By Mark A. McDaniel, Steven F. Maier and Gilles O. Einstein

OLDER ADULTS HAVE shown a strong interest in over-the-counter "brain boosters," many of which are marketed with grand claims touting their benefits. There are sound biochemical reasons for expecting some of these nutrients to be effective. In a review of published research, we found studies showing that some of these substances had robustly enhanced memory in lab animals and occasionally produced impressive improvements in humans as well. Nevertheless, there are numerous questions about the sample sizes in the studies, the generality of the results across different memory tests and populations, and other aspects of the procedures and data. These problems, in conjunction with a general lack of research demonstrating that the results can be replicated, dampen enthusiasm for the effectiveness of these nutrients in substantially arresting or reversing memory loss. Here is an abbreviated summary of the findings for six kinds of nonprescription compounds that are claimed to be memory enhancers and treatments for age-related memory decline.

PHOSPHATIDYLSERINE (PS)

A naturally occurring lipid, PS has been shown to reduce many consequences of aging on the neurons in older rats and mice and to restore their normal memory in a variety of tasks. Research on the impact in humans is limited, though. For older adults with moderate cognitive impairment, PS has produced modest increases in recall of word lists. Positive effects have not been as consistently reported for other memory tests.

CHOLINE COMPOUNDS

Phosphatidylcholine, which is typically administered as lecithin, has not proved effective for improving memory in patients with probable Alzheimer's disease. Research on citicoline is practically nonexistent, but one study reported a robust improvement in story recall for a small sample of normally aging older adults.

PIRACETAM

Developed in 1967, Piracetam has not been approved by the U.S. Food and Drug Administration, but it is sold in Europe and Mexico under several names (including Nootropil and Pirroxil). Animal studies suggest that the drug may improve neural transmission and synaptic activity and also combat age-related deterioration of neuronal membranes. But there is no clear sign of any cognitive benefits in patients with probable Alzheimer's or adults with age-associated memory impairment.

VINPOCETINE

An alkaloid derived from the periwinkle plant, vinpocetine increases blood circulation in the brain. In three studies of older adults with memory problems associated with poor brain circulation or dementia-related disease, vinpocetine produced improvements in performance on cognitive tests measuring attention, concentration and memory.

ACETYL-L-CARNITINE (ALC)

An amino acid included in some "brain power" supplements sold in health food stores, ALC participates in cellular energy production, a process especially important to neurons. Animal studies show that ALC reverses age-related decline in the number of receptor molecules on neuronal membranes. But studies of patients with probable Alzheimer's have reported only nominal advantages for ALC in a range of memory tests.

ANTIOXIDANTS

Antioxidants such as vitamins E and C help to neutralize tissue-damaging free radicals, which become more prevalent with age. But studies have found that vitamin E does not offer significant memory benefits for patients with Alzheimer's or early Parkinson's disease. A combination of vitamins E and C did not improve college students' performance on several cognitive tasks.

A more detailed version of this article appeared as "Brain-Specific Nutrients: A Memory Cure?" by Mark A. McDaniel, Steven F. Maier and Gilles O. Einstein in Psychological Science in the Public Interest (May 2002). (Available at www.psychologicalscience.org/journals/pspi/3_1.html) McDaniel is chair of the department of psychology at the University of New Mexico. Maier is director of the Center for Neuroscience at the University of Colorado at Boulder. Einstein is chair of the department of psychology at Furman University.

BUYERS BEWARE: Researchers have not found convincing evidence that dietary supplements can enhance memory.

CHUCK SAVAGE Corbis

risks, particularly at the typical doses of 120 to 240 milligrams a day. Some complications have been noted, though, including subdural hematomas (clots between the skull and brain) and gastrointestinal problems. As is the case with most plant extracts and medications, ingestion of ginkgo is at times associated with nausea and vomiting. In addition, some users experience increased salivation, decreased appetite, headaches, dizziness, tinnitus (ringing in the ears) and skin rash. Large doses may lead to orthostatic hypotension, a condition of low blood pressure sometimes seen after abrupt postural changes, such as standing up after being seated. Still, the general impression is that the incidence of serious adverse consequences after use of ginkgo is quite low. Also, this incidence may be reduced further if and when optimal individual dose regimens for ginkgo are established.

Because of the differences in experimental designs used to test ginkgo and other treatments, it is difficult to make direct comparisons of efficacy. For instance, on a test involving memory of a brief story, glucose enhanced performance in young adult and healthy elderly subjects by about 30 to 40 percent. In people with Alzheimer's, the improvement on a similar memory test approached 100 percent, with smaller enhancements seen on other measures. The extent of improvement in these experiments is much larger than the 10 to 20 percent gain shown with ginkgo. But most of the experiments testing the effects of glucose have used short-term treatments and compared performance before and afterward, whereas most of the experiments testing ginkgo have used long-term treatments and compared ginkgo-treated subjects with a control group.

Does ginkgo biloba in fact enhance cognitive function? The proof for even a MILD BENEFIT is weak.

But to return to our original question, does ginkgo biloba in fact enhance cognitive function? In general, the reported effects are rather small. The number of experiments is also small, and they are of mixed quality, so the proof for even a mild benefit is weak. In humans, ginkgo may slow cognitive decline during dementia. It is possible that ginkgo's main effects on memory kick in after one dose and are relatively short-lived, but the research literature is so limited that significant issues such as this one cannot be adequately evaluated at this time.

The Bottom Line

GIVEN THE AVAILABLE EVIDENCE, is ginkgo biloba the best therapy for improving memory? Other supplements have been found to influence cognitive function in humans and lab animals [*see box on opposite page*]. Pharmaceuticals such as donepezil can strongly enhance learning and memory in rodents and induce modest though significant improvements in humans. But relatively simple interventions can produce some of the same results. For example, hearing an exciting story apparently releases epinephrine from the adrenal glands into the circulation, thereby enhancing one's memory without any drugs at all. One mechanism by which epinephrine might enhance memory is by liberating glucose from stores in the liver, thereby increasing the circulating glucose available to the brain.

Eating a simple sugar can also improve one's memory. Considerable evidence indicates that glucose administered systemically (to humans by ingestion and to rodents by injection) enhances cognitive performance in young and elderly rats, mice and humans, including people with Alzheimer's. Like most treatments that improve memory, glucose's effects follow a dose-response curve in the shape of an inverted U. Only intermediate doses improve memory; low doses are ineffective, and high doses may actually impair memory.

Establishing direct comparisons of efficacy is vital to identifying which treatments improve cognition the most. This is one of many instances in which further studies of rodents would be useful because of the researchers' ability to control all the variables in the experiment. Only one study has directly compared the effects of ginkgo with those of other treatments. This study showed that the peak enhancement observed with ginkgo was about half of that seen with other drugs. More direct comparisons, in both humans and lab animals, are clearly needed.

We began our survey of the research literature with healthy skepticism but with a commitment to avoid prejudging the findings. We found evidence supporting the view that ginkgo enhances cognitive functions, albeit rather weakly, under some conditions. Our overriding impression, however, is that we do not have enough information to say conclusively whether ginkgo does or does not improve cognition. There are too few experiments on which to base clear recommendations, and most of the studies showing benefits have involved too few subjects.

But there are enough positive findings, perhaps just enough, to sustain our interest in conducting further research on ginkgo. Many years of experience with investigations of new drugs have demonstrated that the initial positive results from studies involving a small number of subjects tend to disappear when the drugs are tested in larger numbers of subjects from diverse populations. So the true test of ginkgo's efficacy lies ahead. **SA**

MORE TO EXPLORE

A more detailed version of this article appeared as "**Ginkgo Biloba: A Cognitive Enhancer?**" by Paul E. Gold, Larry Cahill and Gary L. Wenk in *Psychological Science in the Public Interest*, Vol. 3, No. 1, pages 2–11; May 2002. Available at www.psychologicalscience.org/journals/pspi/3_1.html

More information about ginkgo biloba and other dietary supplements can be found at **dietary-supplements.info.nih.gov/** and **www.cfsan.fda.gov/~dms/supplmnt.html**

The Lowdown on Ginkgo Biloba
IN REVIEW

TESTING YOUR COMPREHENSION

1) What part of the ginkgo tree is used as a dietary supplement?

    a) ground root   c) leaf extract

    b) seeds   d) dried bark

2) Why doesn't the U.S. Food and Drug Administration (FDA) regulate the use of ginkgo biloba?

    a) It is an herbal treatment.

    b) It is naturally derived.

    c) It has never been proven to be harmful.

    d) It is a common drug.

3) Acetylcholine is an example of a(n) _____, a brain chemical that conveys signals between neurons.

    a) amide

    b) drug

    c) selective serotonin reuptake inhibitor

    d) neurotransmitter

4) Which of the following is NOT believed to be an effect of gingko biloba on the brain?

    a) It increases brain circulation.

    b) It stimulates the growth of new synapses between neurons.

    c) It increases the absorption of glucose.

    d) It affects neurotransmitter production and release.

5) Which of the following is NOT a concern associated with the human consumption of ginkgo biloba?

    a) Its effects on cognitive function are not fully understood.

    b) It sometimes causes complications such as blood clots.

    c) It sometimes causes nausea and vomiting.

    d) It causes increased risk of heart disease in animals and thus may do so in humans.

6) Generally speaking, how high is the risk of serious health problems from using the standard dose of ginkgo biloba?

    a) zero

    b) very low

    c) fairly high

    d) very high

7) Which of the following statements describes an observed relationship between glucose ingestion and cognitive functions in humans?

    a) Moderate doses of glucose improve memory.

    b) Low doses of glucose improve memory.

    c) High doses of glucose improve memory.

    d) No amount of glucose affects memory.

8) Why would it be valuable to perform tests on the effects of ginkgo biloba using mice instead of humans?

    a) Mice and humans have identical brain chemistry.

    b) Mice are smaller and therefore require a smaller dose.

    c) Researchers can better control the variables of the experiment.

    d) All of the above

9) Which of the following problems are associated with research into the efficacy of so-called "brain boosters" on human cognition?

    a) small sample sizes

    b) difficulty in generalizing across populations

    c) lack of reproducibility

    d) all of the above

10) Which of the following statements best summarizes the current scientific understanding of the effects of ginkgo biloba on human cognition?

    a) Ginkgo biloba probably has no effect on cognition.

    b) Ginkgo biloba appears to slightly increase cognitive functions, but the effect is small.

    c) Ginkgo biloba appears to slightly decrease cognitive functions, but the effect is small.

    d) Ginkgo biloba appears to increase cognitive functions a lot.

BIOLOGY IN SOCIETY

1) Because ginkgo biloba is an herbal remedy, the FDA does not regulate claims made by its manufacturers. Do you agree with this policy? What is the definition of a drug? In what ways is ginkgo biloba different than what you think of as a "drug?" In what ways is it similar?

2) In order to determine ginkgo biloba's effectiveness as a treatment for Alzheimer's Disease, it is necessary to conduct experiments on people with the disease. Do you see any ethical problems inherent in that? Is it unethical to perform experiments on people who cannot give truly informed consent? If you were the caretaker of an Alzheimer's patient, would you give permission for them to participate in such a study? Why or why not? If you developed Alzheimer's Disease, does it bother you to imagine someone enrolling you in a study?

THINKING ABOUT SCIENCE

1) On page 56, the author describes a selection of some of the most rigorous experiments into the relationship between gingko biloba and cognitive function. One of the criteria used to choose the most rigorous studies is that they be of "placebo-controlled, double-blind design." What does that mean? Describe an experimental design that meets this criteria. Why is that important?

2) On page 59, the author states that gingko biloba research could benefit from studies using rodents because, with mice, experimental variables can be better controlled. What does this mean? Which variables can be better controlled when studying mice than humans? Why is this helpful?

WRITING ABOUT SCIENCE

Ginkgo biloba is just one of many herbal remedies popular within our culture today. Choose another popular herbal remedy. See if you can determine the claims made by its manufacturers. Are the claims consistent? Can you find any studies that support or refute the claim? Do the claims have some other basis, perhaps in folklore medicine?

Testing Your Comprehension Answers:
1c, 2a, 3d, 4b, 5d, 6b, 7a, 8c, 9d, 10b.

THE Last Fish

OVERFISHING HAS SLASHED STOCKS—ESPECIALLY OF LARGE PREDATOR SPECIES—TO AN ALL-TIME LOW WORLDWIDE, ACCORDING TO NEW DATA. IF WE DON'T MANAGE THIS RESOURCE, WE WILL BE LEFT WITH A DIET OF JELLYFISH AND PLANKTON STEW

By Daniel Pauly and Reg Watson

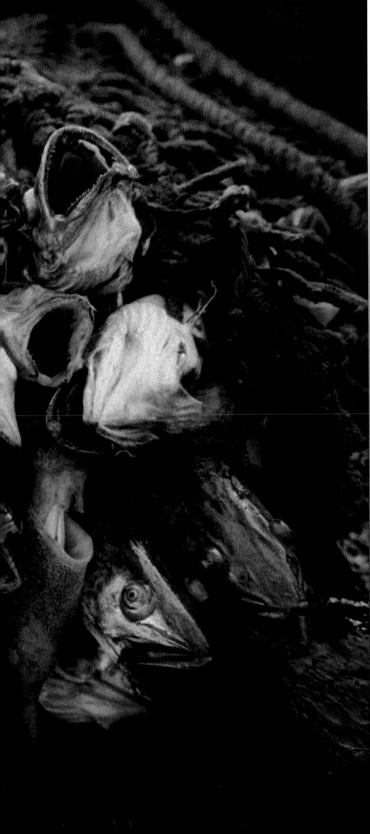

Georges Bank—the patch of relatively shallow ocean just off the coast of Nova Scotia, Canada—used to teem with fish. Writings from the 17th century record that boats were often surrounded by huge schools of cod, salmon, striped bass and sturgeon. Today it is a very different story. Trawlers trailing dredges the size of football fields have literally scraped the bottom clean, harvesting an entire ecosystem—including supporting substrates such as sponges—along with the catch of the day. Farther up the water column, longlines and drift nets are snagging the last sharks, swordfish and tuna. The hauls of these commercially desirable species

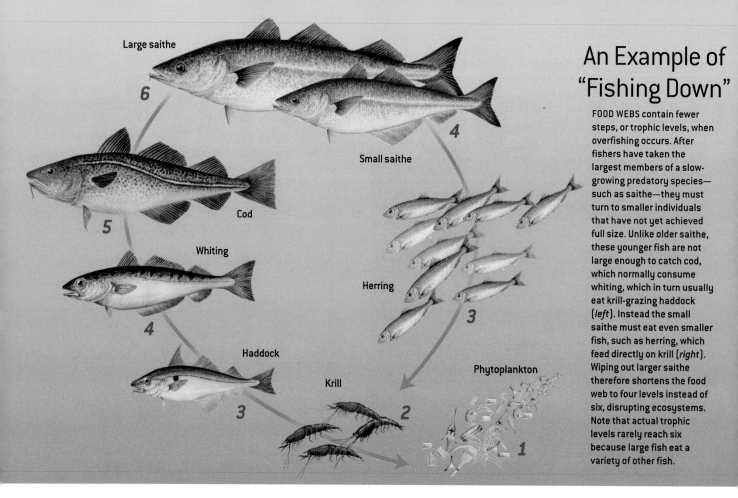

An Example of "Fishing Down"

FOOD WEBS contain fewer steps, or trophic levels, when overfishing occurs. After fishers have taken the largest members of a slow-growing predatory species—such as saithe—they must turn to smaller individuals that have not yet achieved full size. Unlike older saithe, these younger fish are not large enough to catch cod, which normally consume whiting, which in turn usually eat krill-grazing haddock (*left*). Instead the small saithe must eat even smaller fish, such as herring, which feed directly on krill (*right*). Wiping out larger saithe therefore shortens the food web to four levels instead of six, disrupting ecosystems. Note that actual trophic levels rarely reach six because large fish eat a variety of other fish.

Labels in figure: Large saithe, Small saithe, Cod, Whiting, Herring, Haddock, Krill, Phytoplankton

are dwindling, and the sizes of individual fish being taken are getting smaller; a large number are even captured before they have time to mature. The phenomenon is not restricted to the North Atlantic but is occurring across the globe.

Many people are under the mistaken impression that pollution is responsible for declines in marine species. Others may find it hard to believe that a shortage of desirable food fish even exists, because they still notice piles of Chilean sea bass and tuna fillets in their local fish markets. Why is commercial fishing seen as having little if any effect on the species that are being fished? We suspect that this perception persists from an-

other age, when fishing was a matter of wresting sustenance from a hostile sea using tiny boats and simple gear.

Our recent studies demonstrate that we can no longer think of the sea as a bounteous provider whose mysterious depths contain an inexhaustible resource. Over the past several years we have gathered and analyzed data on the world's fisheries, compiling the first comprehensive look at the state of the marine food resource. We have found that some countries, particularly China, have overreported their catches, obscuring a downward trend in fish caught worldwide. In general, fishers must work farther offshore and at greater depths in an effort to keep up with the catches of yesteryear and to try to meet the burgeoning demand for fish. We contend that overfishing and the fishing of these distant stocks are unsustainable practices and are causing the depletion of important species. But it is not too late to implement policies to protect the world's fisheries for future generations.

The Law of the Sea

EXPLAINING HOW THE SEA got into its current state requires relating a bit of history. The ocean used to be a free-for-all, with fleets flying the flags of various countries competing for fish thousands of miles from home. In 1982 the United Nations adopted the Convention on the Law of the Sea, which allows countries bordering the ocean to claim exclusive economic zones reaching 200 nautical miles into open waters. These areas include the highly productive continental shelves of roughly 200 meters in depth where most fish live out their lives.

The convention ended decades—and, in some instances, even

Overview/*Fish Declines*

- New analyses show that fisheries worldwide are in danger of collapsing from overfishing, yet many people still view the ocean as a limitless resource whose bounty humanity has just begun to tap.
- Overfishing results from booms in human populations, increases in the demand for fish as a nutritious food, improvements in commercial fishing technology, and global and national policies that fail to encourage the sustainable management of fisheries.
- Solutions to the problem include banning fishing gear such as dredges that damage ecosystems; establishing marine reserves to allow fisheries to recover; and abolishing government subsidies that keep too many boats on the seas chasing too few fish.

centuries—of fighting over coastal fishing grounds, but it placed the responsibility for managing marine fisheries squarely on maritime countries. Unfortunately, we cannot point to any example of a nation that has stepped up to its duties in this regard.

The U.S. and Canadian governments have subsidized the growth of domestic fishing fleets to supplant those of now excluded foreign countries. Canada, for instance, built new offshore fleets to replace those of foreign nations pushed out by the convention, effectively substituting foreign boats with even larger fleets of more modern vessels that fish year-round on the same stocks that the domestic, inshore fleet was already targeting. In an effort to ensure that there is no opportunity for foreign fleets to fish the excess allotment—as provided for in the convention—these nations have also begun to fish more extensively than they would have otherwise. And some states, such as those in West Africa, have been pressured by others to accept agreements that allow foreign fleets to fish their waters, as sanctioned by the convention. The end result has been more fishing than ever, because foreign fleets have no incentive to preserve local marine resources long-term—and, in fact, are subsidized by their own countries to garner as much fish as they can.

The expansion made possible by the Convention on the Law of the Sea and technological improvements in commercial fishing gear (such as acoustic fish finders) temporarily boosted fish catches. But by the late 1980s the upward trend began to reverse, despite overreporting by China, which, in order to meet politically driven "productivity increases," was stating that it was taking nearly twice the amount of fish that it actually was.

In 2001 we presented a statistical model that allowed us to examine where catches differed significantly from those taken from similarly productive waters at the same depths and latitudes elsewhere in the world. The figures from Chinese waters—about 1 percent of the world's oceans—were much higher than predicted, accounting for more than 40 percent of the deviations from the statistical model. When we readjusted the worldwide fisheries data for China's misrepresentations, we concluded that world fish landings have been declining slowly since the late 1980s, by about 700,000 metric tons a year. China's overreporting skewed global fisheries statistics so significantly because of the country's large size and the degree of its overreporting. Other nations also submit inaccurate fisheries statistics—with a few overreporting their catches and most underreporting them—but those numbers tend to cancel one another out.

Nations gather statistics on fish landings in a variety of ways, including surveys, censuses and logbooks. In some countries, such as China, these data are forwarded to regional offices and on up through the government hierarchy until they arrive at the national offices. At each step, officials may manipulate the statistics to meet mandatory production targets. Other countries have systems for cross-checking the fish landings against import/export data and information on local consumption.

The most persuasive evidence, in our opinion, that fishing is wreaking havoc on marine ecosystems is the phenomenon that one of us (Pauly) has dubbed "fishing down the food web." This describes what occurs when fishers deplete large preda-

Hot Spots of Overfishing

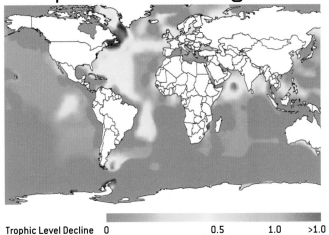

Trophic Level Decline   0   0.5   1.0   >1.0

OVERFISHING caused the complexity of the food chains in important fisheries to drop by more than one trophic level between the years 1950 and 2000. The open ocean usually has few fish.

tor fish at the top of the food chain, such as tuna and swordfish, until they become rare, and then begin to target smaller species that would usually be eaten by the large fish [see illustration on opposite page].

Fishing Down

THE POSITION A PARTICULAR ANIMAL occupies in the strata of a food web is determined by its size, the anatomy of its mouthparts and its feeding preferences. The various layers of the food web, called trophic levels, are ranked according to how many steps they are removed from the primary producers at the base of the web, which generally consists of phytoplanktonic algae. These microscopic organisms are assigned a trophic level (TL) of 1.

Phytoplankton are grazed mostly by small zooplankton—mainly tiny crustaceans of between 0.5 and two millimeters in size, both of which thus have a TL of 2. (This size hierarchy stands in stark contrast to terrestrial food chains, in which herbivores are often very large; consider moose or elephants, for instance.) TL 3 consists of small fishes between 20 and 50 cen-

THE AUTHORS

DANIEL PAULY and REG WATSON are fisheries researchers at the Sea Around Us Project in Vancouver, where Pauly is the principal investigator and Watson is a senior scientist. The project, which was initiated and funded by the Pew Charitable Trusts, is based at the Fisheries Center at the University of British Columbia and is devoted to studying the impact of fishing on marine ecosystems. Pauly's early career centered on formulating new approaches for fisheries research and management in tropical developing countries. He has designed software programs for evaluating fish stocks and initiated FishBase, the online encyclopedia of fishes of the world. Watson's interests include fisheries modeling, data visualization and computer mapping. His current research focuses on mapping the effects of global fisheries, modeling underwater visual census techniques and using computer simulations to optimize fisheries.

NINA FINKEL

THE AUDUBON Seafood Wallet Card

choose your seafood wisely

Farmed mussels and clams
Alaska salmon
Mahimahi, *troll caught*
Crawfish
Alaska halibut
Dungeness crab

Fold along dotted line — Tilapia, *U.S. farmed*

Yellowfin, bigeye, albacore tuna, *pole/troll caught*
Catfish
Striped bass
Mahimahi, *longline caught*
Pacific cod
Pacific flounders and soles
Rainbow trout
American (Maine*) lobster
Squid (calamari)
Ahi Tuna (yellowfin and bigeye tuna steak)
Canned tuna

Fold along dotted line — Swordfish

Atlantic cod
Groupers
Shrimp
Atlantic flounders and soles
Monkfish
Sharks
Farmed salmon (including Atlantic)
Orange roughy
Snappers
Chilean seabass (toothfish)
Atlantic halibut

* frequently called "Maine" lobster but not always from Maine

POPULAR FISH—including many of the fillets and steaks that can be found in piles at fish markets (*above*)—have been decimated by overfishing. Fishers must use increasingly complex technology and fish farther offshore and at greater depths to catch such fish. The National Audubon Society and other organizations have issued wallet cards (*right*) so that consumers can avoid overfished species (*red*) or those whose status is cause for concern (*yellow*). The entire card can be downloaded at www.audubon.org/campaign/lo/seafood/cards.html

timeters in length, such as sardines, herring and anchovies. These small pelagic fishes live in open waters and usually consume a variable mix of phytoplankton and both herbivorous and carnivorous zooplankton. They are caught in enormous quantities by fisheries: 41 million metric tons were landed in 2000, a number that corresponds to 49 percent of the reported global marine fish catch. Most are either destined for human consumption, such as canned sardines, or reduced to fish meal and oil to serve as feed for chickens, pigs and farmed salmon or other carnivorous fish.

The typical table fish—the cod, snapper, tuna and halibut that restaurants serve whole or as steaks or fillets—are predators of the small pelagics and other small fishes and invertebrates; they tend to have a TL of between 3.5 and 4.5. (Their TLs are not whole numbers because they can consume prey on several trophic levels.)

The increased popularity in the U.S. of such fish as nutritious foods has undoubtedly contributed to the decline in their stocks. We suggest that the health and sustainability of fisheries can be assessed by monitoring the trends of average TLs. When those numbers begin to drop, it indicates that fishers are relying on ever smaller fish and that stocks of the larger predatory fish are beginning to collapse.

In 1998 we presented the first evidence that "fishing down" was already occurring in some fishing grounds, particularly in the North Atlantic, off the Patagonian coast of South America and nearby Antarctica, in the Arabian Sea, and around parts of Africa and Australia. These areas experienced TL declines of 1 or greater between 1950 and 2000, according to our calculations [*see map on preceding page*]. Off the west coast of New-

foundland, for instance, the average TL went from a maximum of 3.65 in 1957 to 2.6 in 2000. Average sizes of fish landed in those regions dropped by one meter during that period.

Our conclusions are based on an analysis of the global database of marine fish landings that is created and maintained by the U.N. Food and Agriculture Organization, which is in turn derived from data provided by member countries. Because this data set has problems—such as overreporting and the lumping of various species into a category called "mixed"—we had to incorporate information on the global distribution of fishes from FishBase, the online encyclopedia of fishes pioneered by Pauly, as well as information on the fishing patterns and access rights of countries reporting catches.

Research by some other groups—notably those led by Jeremy B. C. Jackson of the Scripps Institution of Oceanography in San Diego and Ransom A. Myers of Dalhousie University in Halifax—suggests that our results, dire as they might seem, in fact underestimate the seriousness of the effects that marine fisheries have on their underlying resources. Jackson and his colleagues have shown that massive declines in populations of marine mammals, turtles and large fishes occurred along all coastlines where people lived long before the post–World War II period we examined. The extent of these depletions was not recognized until recently because biologists did not consult historians or collaborate with archaeologists, who study evidence of fish consumption in middens (ancient trash dumps).

Myers and his co-workers used data from a wide range of fisheries throughout the world to demonstrate that industrial fleets generally take only a few decades to reduce the biomass of a previously unfished stock by a factor of 10. Because it often takes much longer for a regulatory regime to be established to manage a marine resource, the sustainability levels set are most likely to be based on numbers that already reflect population declines. Myers's group documents this process particularly well for the Japanese longline fishery, which in 1952 burst out of the small area around Japan—to which it was confined until the end of the Korean War—and expanded across the Pacific and into the Atlantic and Indian oceans. The expansion decimated tuna populations worldwide. Indeed, Myers and his colleague Boris Worm recently reported that the world's oceans have lost 90 percent of large predatory fish.

Changing the Future

WHAT CAN BE DONE? Many believe that fish farming will relieve the pressure on stocks, but it can do so only if the farmed organisms do not consume fish meal. (Mussels, clams and tilapia, an herbivorous fish, can be farmed without fish meal.) When fish are fed fish meal, as in the case of salmon and various carnivores, farming makes the problem worse, turning small pelagics—including fish that are otherwise perfectly fit for human consumption, such as herring, sardines, anchovies and mackerels—into animal fodder. In fact, salmon farms consume more fish than they produce: it can take three pounds of fish meal to yield one pound of salmon.

One approach to resolving the difficulties now besetting the world's fisheries is ecosystem-based management, which would seek to maintain—or, where necessary, reestablish—the structure and function of the ecosystems within which fisheries are embedded. This would involve considering the food requirements of key species in ecosystems (notably those of marine mammals), phasing out fishing gear that destroys the sea bottom, and implementing marine reserves, or "no-take zones," to mitigate the effects of fishing. Such strategies are compatible with the set of reforms that have been proposed for years by various fisheries scientists and economists: radically reducing global fleet capacity; abolishing government subsidies that keep otherwise unprofitable fishing fleets afloat; and strictly enforcing restrictions on gear that harm habitats or that capture "bycatch," species that will ultimately be thrown away.

Creating no-take zones will be key to preserving the world's fisheries. Some refuges should be close to shore, to protect coastal species; others must be large and offshore, to shield

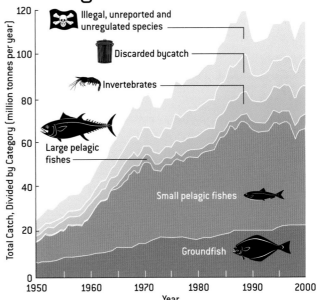

Catching More Fish

AMOUNT OF FISH LANDED has more than quintupled over the past 50 years. As the world's population has grown, commercial fishing technology has advanced, and demand for fish in some countries has surged.

oceanic fishes. No-take zones now exist, but they are small and scattered. Indeed, the total area protected from any form of fishing constitutes a mere 0.01 percent of the ocean surface. Reserves are now viewed by fishers—and even by governments—as necessary concessions to conservationist pressure, but they must become management tools for protecting exploited species from overfishing.

A major goal should be to conserve species that once maintained themselves at deeper depths and farther offshore, before fishers developed improved gear for going after them. This type of fishing is similar to a nonrenewable mining operation because fishes are very vulnerable, typically long-lived, and have very low productivity in the dark, cold depths. These measures would enable fisheries, for the first time, to become sustainable. ⓢⒶ

MORE TO EXPLORE

Effect of Aquaculture on World Fish Supplies. Rosamond L. Naylor, Rebecca J. Goldburg, Jurgenne H. Primavera, Nils Kautsky, Malcolm C. M. Beveridge, Jason Clay, Carl Folke, Jane Lubchenco, Harold Mooney and Max Troell in *Nature*, Vol. 405, pages 1017–1024; June 29, 2000.

Historical Overfishing and the Recent Collapse of Coastal Ecosystems. Jeremy B. C. Jackson et al. in *Science*, Vol. 293, pages 629–638; July 27, 2001.

Systematic Distortion in World Fisheries Catch Trends. Reg Watson and Daniel Pauly in *Nature*, Vol. 414, pages 534–536; November 29, 2001.

In a Perfect Ocean: The State of Fisheries and Ecosystems in the North Atlantic Ocean. Daniel Pauly and Jay Maclean. Island Press, 2003.

Rapid Worldwide Depletion of Predatory Fish Communities. Ransom A. Myers and Boris Worm in *Nature*, Vol. 423, pages 280–283; May 15, 2003.

More information on the state of world fisheries can be found on the Web sites of the Sea Around Us Project at **www.saup.fisheries.ubc.ca** and of FishBase at **www.fishbase.org**

NINA FINKEL (*graph*) AND CLEO VILETT (*illustrations*)

Counting the Last Fish
IN REVIEW

TESTING YOUR COMPREHENSION

1) Why does trawling have a disproportionately large effect on the health of fisheries?
   a) It introduces artificial pollutants.
   b) It removes all species in an area, disrupting the entire ecosystem.
   c) It almost entirely wipes out the lowest trophic levels of the ecosystem.
   d) It reduces the amount of sunlight, disrupting the function of the ecosystem's producers.

2) Which of the following statements is true concerning commercial fishing today?
   a) The size of hauls is decreasing.
   b) The sizes of the fish being caught are decreasing.
   c) More immature fish are being caught.
   d) All of the above

3) After the United Nations' adoption of the 1982 Convention on the Law of the Sea, who controls coastal fisheries?
   a) each maritime country controls the ocean 200 nautical miles off its shore
   b) the United Nations
   c) nobody; any country is allowed to fish anywhere in the ocean
   d) the United States

4) What does trophic level 1 represent?
   a) photosynthetic producers, primarily algae
   b) primary consumers
   c) secondary consumers
   d) humans

5) In general, how do fish change as the trophic level increases?
   a) Fish at higher trophic levels tend to be larger.
   b) Fish at higher trophic levels consume organisms at the lower trophic levels.
   c) Fish at higher trophic levels have smaller population numbers than fish at lower trophic levels.
   d) All of the above

6) Generally speaking, how does overfishing affect food webs?
   a) It increases the number of connections in the web.
   b) It decreases the number of connections in the web.
   c) It decreases the number of trophic levels.
   d) It increases the number of trophic levels.

7) According to the statistical model presented by the authors of this article, how have fish harvests changed?
   a) They have been increasing slowly for about 20 years.
   b) They have been declining slowly for about 20 years.
   c) They have been increasing slowly for about 200 years.
   d) They have been declining slowly for about 200 years.

8) The oceans around which part of the U.S. have been most largely affected by overfishing?
   a) the Northeast
   b) the Southeast
   c) the Gulf of Mexico
   d) the Northwest